LABORATORY MANUAL
for
Digital Electronics through Project Analysis

LABORATORY MANUAL
for
Digital Electronics through Project Analysis

RONALD A. REIS
Los Angeles Trade Technical College

PRENTICE HALL
Englewood Cliffs, NJ 07632

Library of Congress Cataloguing-in-Publication Data

Reis, Ronald A.
    Laboratory manual for digital electronics through project analysis
    / Ronald A. Reis.
      p. cm.
    ISBN 0-675-21254-5 (pbk.)
    1. Digital electronics—Laboratory manuals. I. Title.
TK7868.D5R46 1991
621.381—dc20

90-19987
CIP

Cover photo courtesy of Hughes Aircraft Company

Editor: David Garza
Developmental Editor: Carol Hinklin Robison
Production Editor: Sheryl Glicker Langner
Art Coordinator: Lorraine Woost
Cover Designer: Russ Maselli

This book was set in Times Roman.

 © 1991 by Prentice-Hall, Inc.
A Simon & Schuster Company
Englewood Cliffs, New Jersey 07632

Printed in the United States of America

10 9 8 7 6 5 4 3 2

ISBN 0-675-21254-5

Prentice-Hall International (UK) Limited, *London*
Prentice-Hall of Australia Pty. Limited, *Sydney*
Prentice-Hall Canada Inc., *Toronto*
Prentice-Hall Hispanoamericana, S.A., *Mexico*
Prentice-Hall of India Private Limited, *New Delhi*
Prentice-Hall of Japan, Inc., *Tokyo*
Simon & Schuster Asia Pte. Ltd., *Singapore*
Editora Prentice-Hall do Brasil, Ltda., *Rio de Janeiro*

To the memory of our "Uncle Charlie" Hershfield,
a fine professor who knew the real from the false
and the class from the crass—
you will be missed.

Contents

EXPERIMENTS

APPENDIXES

LABORATORY MANUAL
for
Digital Electronics through Project Analysis

Introduction

A laboratory manual provides a student with the opportunity to prove out the theory that he or she has been learning in the classroom. In doing so, a good laboratory manual should exhibit four prominent characteristics:

1. The manual should be easy to use. The student (and instructor) should not have to spend an inordinate amount of time just trying to figure out what to do or how to do it. Through good organization and layout, the laboratory experiments should allow the student to proceed in a logical, clear, and straightforward manner—with little assistance from others.
2. The manual should excite and challenge the student. The experiments should draw the student in by presenting practical, doable, and, yes, even fun circuits.
3. The manual should be comprehensive; that is, it should cover all aspects of the subject.
4. The manual must contain experiments that work. I have personally constructed every circuit presented here and have done everything you as a student are asked to do. If you proceed in a logical manner, think before you act, and check over each circuit thoroughly before applying power, your circuit will work. I promise.

   I have tried in this laboratory manual to meet the above four objectives. In addition, I want to point out some other features of your manual:

- Note the Experiment Checklist immediately following this Introduction. Use it to keep track of your progress.
- The Master List of Equipment and Materials, which follows the checklist, spells out exactly what you will need in order to complete all experiments. Keep in mind that many of the circuits presented are "miniprojects" in themselves—hence the rather extensive materials list.
- In the section titled The Digital Logic Pulser and Digital Logic Probe: Indispensable Test Instruments for Digital Electronics, we discuss operation of these two very important pieces of equipment. While Radio Shack models are described, what is said applies to any digital logic pulser and probe.
- Since all of your experimenting is to be done on solderless circuit board, the section on Using Solderless Circuit Board is "must" reading. Again, while one type of commercial board is explained, keep in mind that all boards are similar in design.
- Safety, for both you and your circuits, should be of the utmost importance. In the section titled Safety in the Laboratory we look at how you can protect yourself and the

digital circuits you are working on. Read this section carefully; then take the short Safety Quiz.

The bulk of this laboratory manual, of course, consists of experiments—22 in all. Actually, each experiment can be considered three experiments in one. In addition to a list of objectives, an equipment and components list, background information, summary, and review questions, each experiment (except for the first) is made up of three parts:

Part 1, Circuit Fundamentals, explores the subject at an introductory level, in some cases using discrete components.

Part 2, Further Investigation, expands on the basics, explaining advanced chip and circuit features.

Part 3, Circuit Applications, presents one, and in some cases two, ''projectlike'' circuits illustrating the subject under discussion.

The 22 experiments in this laboratory manual, taken together, cover the field of digital electronics. However, it should be pointed out that the *Instructor's Resource Guide* contains additional experiments that may be photocopied and distributed to students. Ask your instructor for details.

In addition, note that this laboratory manual contains four appendixes. Appendix A, IC Pin Diagrams, presents the pin configurations for all the ICs used in the manual. Appendix B, Data Manual Sources, lists 15 companies that will provide, often just for the asking, data manuals on their solid-state products. Appendix C, Troubleshooting: A 10-Point Checklist, simply summarizes the 10 most obvious troubleshooting tips. Appendix D, Equipment You Can Build: A 5- and 9-V Power Supply and a 555 Clock Generator, provides schematic diagrams which will enable you to construct these two most necessary pieces of electronic equipment.

Finally, while this laboratory manual was written to go with the *Digital Electronics Through Project Analysis* text, it can be used quite well with almost any digital textbook.

Best of luck, and welcome to the fascinating world of digital electronics.

EXPERIMENT CHECKLIST

Keeping track of your progress helps you to maintain that progress. In the columns below, record the date you completed or turned in the experiment, the date it was graded or returned, and the grade you received.

Experiment	Date Completed (or Turned In)	Date Graded (or Returned)	Grade Received
1.			
2.			
3.			
4.			
5.			
6.			
7.			
8.			
9.			
10.			
11.			
12.			
13.			

14. _____ _____ _____

15. _____ _____ _____

16. _____ _____ _____

17 _____ _____ _____

18. _____ _____ _____

19. _____ _____ _____

20. _____ _____ _____

21. _____ _____ _____

22. _____ _____ _____

MASTER LIST OF EQUIPMENT AND MATERIALS

Below is the master list of equipment and materials (electronic components) needed to complete all the experiments in this manual. Though the list is extensive, with a little shortcutting and combining, you can certainly pare it down a bit. For example, with regard to equipment, two items can be built from plans presented in Appendix D: the 5- and 9-V Power Supply and the 555 Clock Generator. Two pieces of test equipment, the function generator and the radio-frequency (RF) generator, are used for only one experiment each. Furthermore, if you don't have a power supply with -12 and $+12$ V, two center-tapped series-connected 9-V batteries will do. In addition, while a dual-channel oscilloscope is definitely preferable, you can get by with the single-channel version. Finally, if you already possess one of the many excellent digital trainers on the market, your equipment list is practically already at hand.

When it comes to materials, some combining or eliminating is also possible. If you don't have green or yellow LEDs, simply use red ones. The liquid-crystal seven-segment display is used as part of only one experiment (Experiment 8)—if you can't find one, just skip that activity. Many of the capacitor values are not at all critical; besides, you can usually series/parallel on-hand values to get what you want. The same can be said of resistors, of course. Finally, all switches, even the push buttons, can be replaced by simple jumper wires in a pinch.

Equipment

1   5- and 9-V power supply (see Appendix D)
1   -12- and $+12$-V power supply
1   digital logic probe (Radio Shack No. 22-303 or equivalent)
1   digital logic pulser (Radio Shack No. 22-304 or equivalent)
1   555 clock generator (see Appendix D)
1   function generator
1   radio-frequency (RF) generator
1   voltmeter (preferably digital)
1   dual-channel oscilloscope
1   solderless circuit board ($2'' \times 6''$ minimum)

Materials

Integrated Circuits TTL

1   7400 quad two-input NAND gate
1   7402 quad two-input NOR gate
1   7404 hex inverter
1   7408 quad two-input AND gate
1   7432 quad two-input OR gate
1   7442 BCD-to-decimal decoder
1   7447 BCD-to-seven-segment decoder/driver
2   7473 dual master-slave J-K flip-flops
1   7474 dual positive edge-triggered D flip-flop

1	7475 quad latch
1	7483 4-bit full adder
2	7485 4-bit magnitude comparators
2	7486 quad two-input exclusive-OR gates
1	7489 64-bit (16 × 4) RAM
2	7490 decade counters
1	7493 binary counter
1	74121 monostable multivibrator
1	74LS132 quad two-input NAND Schmitt trigger
1	74147 10-line-to-4-line priority encoder
1	74148 priority encoder
1	74151 data selector/multiplexer
1	74153 dual 4-line-to-1-line multiplexer
1	74154 4-line-to-16-line decoder/demultiplexer
1	74155 2-line-to-4-line demultiplexer
1	74193 synchronous up/down 4-bit binary counter
2	74194 4-bit bidirectional universal shift registers
1	74LS266 quad two-input exclusive-NOR gate

Integrated Circuits, Linear

1	555 timer
1	741 operational amplifier
1	ADC0804 8-bit A/D converter
1	DAC0808 8-bit D/A converter

Transistors and Other Solid-State Components

1	1N4001 1-A 50-V diode
1	2N3904 NPN transistor
1	2N3906 PNP transistor
1	C106B1 silicon-controlled rectifier (SCR)

Light-Emitting Diodes (LEDs) and Liquid-Crystal Displays

20	LEDs (red)
2	LEDs (green)
1	LED (yellow)
1	common-anode seven-segment LED display
1	liquid-crystal seven-segment display

Capacitors (All Rated at 10 V or Higher)

1	100 pF
2	0.01 μF
1	0.047 μF
2	0.1 μF
2	0.47 μF
1	1 μF
2	4.7 μF
1	10 μF
2	47 μF
1	100 μF
2	150 μF
1	470 μF
1	1000 μF

Resistors (All Fixed Resistors Are ¼ W)

1	47 Ω
1	100 Ω
8	220 Ω
1	330 Ω
1	470 Ω

16	1 kΩ
4	2 kΩ
3	4.7 kΩ
4	10 kΩ
1	12 kΩ
0	15 kΩ
1	18 kΩ
1	22 kΩ
1	51 kΩ
1	100 kΩ
7	1 MΩ
1	2.2 MΩ
1	1-kΩ potentiometer
1	100-kΩ potentiometer
1	100-kΩ potentiometer (10-turn trimmer pot if possible)
1	1-MΩ potentiometer
1	photoresistor (photocell: Radio Shack 276-116 or equivalent)

Switches

3	N.O. push-button switches
1	N.C. push-button switch
2	8-position DIP switches
1	SPST slide switch

Miscellaneous Materials

1	speaker, 8 Ω
1	package of jumper, or hookup, wires (22–24 gauge, approximately 40 pieces of assorted lengths)

A kit of the above components is available from:

EKI Incorporated
16631 Noyes Avenue
Irvine, California 92714
(800) 453-1708

THE DIGITAL LOGIC PULSER AND DIGITAL LOGIC PROBE

The digital logic pulser and digital logic probe are simple but essential instruments employed to test logic circuits. The former is used to deliver a single pulse or pulse train to an input of the circuit under test. The latter is used to display a LOW, HIGH, or pulse condition at a circuit output. Commercial digital logic pulsers and probes, like those shown in Figure 1, are available for under $20 each. No electronics technician should be without them. Both instruments are required to complete the experiments in this manual. Let's take a moment to examine each one.

A typical digital logic pulser can provide a single pulse or a continuous signal (pulse train) to TTL or CMOS circuitry. The instrument usually gets its power from the circuit under test. A slide switch is used to place the pulser in the pulse or continuous signal mode. If the switch is put in the pulse mode, a short 5-μs pulse (Figure 2a) will be generated every time a "finger" push button is pressed. When the switch is in the continuous mode, a 5-Hz pulse train is generated as the push button is pressed and held down (Figure 2b). Green and red LEDs are used to indicate normal and overload operation, respectively.

A typical digital logic probe will test TTL or CMOS circuits by displaying LOW, HIGH, and pulse conditions with different colored LEDs and audible signals when the probe tip is applied to various voltage test points. It too gets its power from the circuit under test.

The digital logic pulser and digital logic probe are often used together to test a digital circuit. For example, if you wish to test an AND gate for proper operation, you might make connections as shown in Figure 3. The output should follow the input.

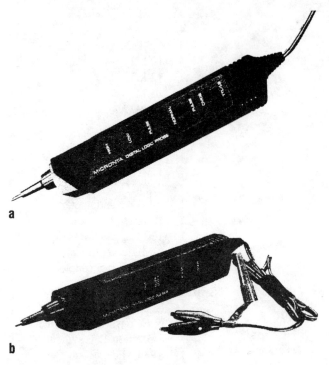

FIGURE 1   (a) Digital logic pulser and (b) digital logic probe. *(Courtesy of Radio Shack: A Division of Tandy Corporation.)*

FIGURE 2   Digital logic pulser
in (a) pulse mode and
(b) continuous mode.

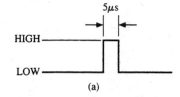

(a)

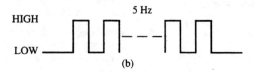

(b)

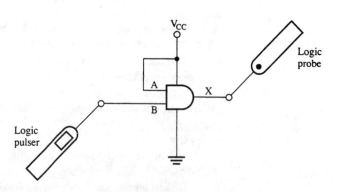

FIGURE 3   Testing an AND gate with the pulser and the probe.

Digital Logic Pulser Setup

To test any digital circuit, the digital logic pulser is set up as follows:

1. The black lead of the pulser is connected to the negative (−) and the red lead to the positive (+) terminal of the power source of the circuit under test.
2. The pulse/continuous switch is set to the pulse or continuous position.
3. The tip pin is touched to an input of the circuit under test.
4. The push button is pressed.

Digital Logic Probe Setup

If the digital logic probe is to be used, it is set up as follows:

1. The black lead of the logic probe is connected to the negative (−) and the red lead to the positive (+) terminal of the power source of the circuit under test.
2. The device-type switch is set to TTL or CMOS, depending on which "family" of logic devices is going to be tested.
3. The mode switch is set to either normal or pulse. In the normal mode, the logic probe will indicate the DC logic state. In the pulse mode, the logic probe will indicate the logic level of the pulse train.
4. The exposed metal tip of the logic probe is touched to a circuit point to be tested. A LOW-level output is indicated by a lit green LED and a low tone. A HIGH-level output is indicated by a lit red LED and a high tone. A pulse is indicated by a blinking yellow LED (when the probe switch is in the pulse position). If no LED is lit and no sound is emitted, the level is either between a LOW and a HIGH or the circuit is "open."

After obtaining your digital logic pulser and digital logic probe, read and study their accompanying instruction manuals. Know your instruments thoroughly before using them. And remember, with proper care both devices should last a lifetime and provide trouble-free and effective operation.

USING SOLDERLESS CIRCUIT BOARD

All the experiments in this laboratory manual are to be performed on what is generally known as *solderless circuit board*, a breadboarding material designed specifically to accommodate integrated circuits. Let's examine what solderless circuit board is and how it is used.

Solderless Circuit Board: What Is It?

Solderless circuit board is a plastic breadboard with hundreds of holes in it (Figure 4). Such boards come in many different sizes—¼ × 2 × 6 inches being the most popular. They can be snapped or butted together to increase the working surface. The holes are placed in a rectangular grid at regular intervals, 0.10 inches apart. Some boards have number and letter designations for each column and a row of holes to aid in component lead and wire placement.

Each hole has a tiny metal lug inside. The lugs are connected underneath in small and large groups. The usual pattern is to connect five lugs in a vertical column of small groups and 25 to 40 or more lugs in large horizontal groups, known as *buses* (Figure 4).

The system works like this: Two or more component leads or wires inserted in the same group are "shorted" together by the lug-connecting strip underneath. The wire (or component lead) can usually be pushed in or pulled out with your fingers, although a needle-nose pliers is handy.

One important feature in every solderless circuit board, regardless of the manufacturer, is the center channel. Integrated circuits are straddled across the channel so that one row of IC pins does not "short" to the row on the opposite side.

In addition to the solderless circuit board itself, the only other material used with this system is hookup wire. You can buy a kit of insulated precut, prestripped (¼ inches

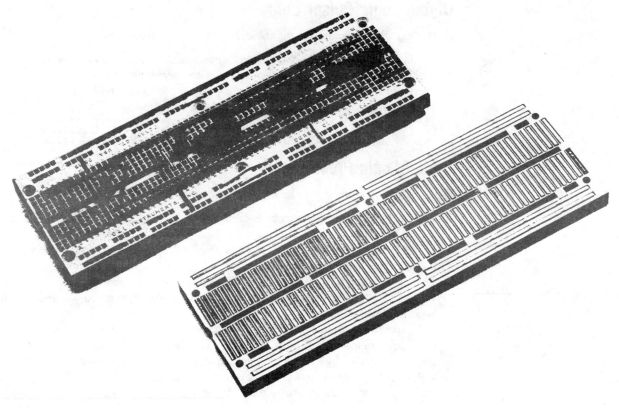

FIGURE 4 Solderless circuit board. *(From Ronald Reis, Electronic Project Design and Fabrication, Merrill, 1989.)*

from the end), single-strand 22-gauge wire from a number of distributors. Or you can use commonly available 22- to 30-gauge telephone wire and do the cutting and stripping yourself.

Solderless Circuit Board: How Is It Used?

Circuit assembly using solderless circuit board, as shown in Figure 5, is very quick and easy. Note, for example, how resistor R_1 is connected between pins 7 and 8 on the 555 IC by simply inserting each lead in the vertical column of holes aligned with those respective pins. If two components must come together at a location other than an IC pin, simply pick a free row of pins and insert. This was done with the junction of the LED and resistor R_3 in Figure 5. Where many component leads or wires come together at a single point, the bus strips can be used.

In many respects, actual layout of the circuit follows the schematic drawing. As with the schematic, the supply bus is at the top, the signal bus in the middle, and the ground bus on the bottom.

When breadboarding a circuit, keep the following tips in mind:

- Keep wires as close to the board as possible. Avoid the "rat's nest" approach. A neat layout is always much easier to trace.
- Keep component leads short for the same reason.
- It is easy to misalign IC pins, component leads, and wires. Check carefully to be sure you haven't inadvertently moved a connection over a hole or two.
- It is very easy to tuck IC pins under the chips without knowing it. From the outside it looks fine; the pin seems to be inserted into the hole. Yet it may be bent underneath the integrated circuit. To avoid this problem, press evenly on all pins at the same time.
- For component leads that are too big to insert into the board, simply solder a smaller wire to the component lead and press its free end into the board.

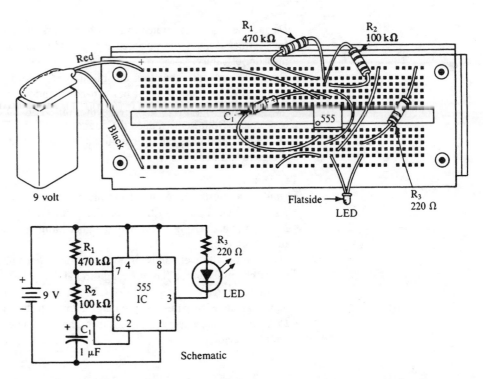

FIGURE 5 Circuit assembly using solderless circuit board. *(From Ronald Reis, Electronic Project Design and Fabrication, Merrill, 1989.)*

- When possible, use different colored wires for different electrical functions. For example, use black wire for ground connections and red wire for positive voltage lines.

SAFETY IN THE LABORATORY

To protect both yourself and the circuit you build, you need to become thoroughly familiar with basic electronics laboratory safety procedures. Here we'll look at human safety (your protection) first, then examine what you can do to protect the circuit you're working on, particularly from static electricity.

Human Safety

Your safety begins by maintaining a safe work environment. Doing so isn't complicated; it just involves following four simple rules:

1. Keep your work area neat and orderly. Being able to see what you have and what you are doing will eliminate many problems.
2. Be alert and attentive at all times. Distractions of any kind can only leave you open to risk.
3. Know the correct safety procedures. If you are not sure how to operate a piece of test equipment, a tool, or any machinery safely, ask your instructor. There is a safe way to do everything. Learn what it is—and use it.
4. If an accident does occur, know what action to take. Be familiar with the basic first-aid measures that could mean the difference between a quick recovery and a serious medical complication.

In addition to maintaining a safe work environment, observing the following do's and don'ts while working with electricity will increase your safety margin considerably:

The Do's
- *Do* work with one hand behind your back while testing live circuits. In that way, if you complete the path for current flow, at least it won't be through your heart.

- *Do* use an isolation transformer while working on AC-powered equipment. The transformer isolates the powered equipment from the power source, adding a strong measure of safety.
- *Do* make sure all capacitors are discharged before troubleshooting begins. Use an insulated screwdriver to short out capacitor leads.
- *Do* use three-conductor grounded line cords and polarized plugs with AC-operated equipment. Both items reduce the danger from a short-circuited chassis.
- *Do* keep your fingers out of live circuits. Test all circuits with a voltmeter or specially designed test lamps.

The Don'ts

- *Don't* install or remove any electronic components while the circuit is connected to a power source. Following this procedure will protect you as well as the component.
- *Don't* overfuse. Using a fuse with a higher rating than is recommended only defeats the fuse's purpose.
- *Don't* work with wet hands. If body resistance goes down, current flow goes up. Even sweat can moisten the hands enough to cause excessive current flow.
- *Don't* cut wires carrying electricity. Remember, it isn't just the AC line cord that is lethal. Assume all wires carry enough current to harm you.
- *Don't* disconnect electrical devices from the wall outlet by pulling on the line cord; pull the plug handle.

Circuit Safety

Electronic circuits can undoubtedly do you harm, but you can in turn harm them, especially with static electricity. Static electricity is a stationary charge. As such, it is a potential (voltage) waiting to go somewhere. If that charge gets to a sensitive circuit, it can destroy the circuit. In the normal course of your work, it is possible to build up a static charge of more than 50,000 V. But today's solid-state circuits don't need anything like that kind of voltage to be rendered useless. In some cases, less than a few volts of static electricity is all that it takes.

The most common way of creating a static charge is through friction, where electrons are torn off one surface to accumulate on another. Any time objects are rubbed together, friction is created. And if the substances are dissimilar materials, the accumulated charge can be considerable. If one or both of the elements is nonconducting, the charge remains for a long time. It has no place to go. On the other hand, if both materials are good conductors, the charges will leak off quickly and little danger exists. The solution then is either to keep charged insulating materials away from sensitive circuits or to make sure the charge is bled away before it can do any damage.

Any CMOS (complementary metal-oxide semiconductor) integrated circuit can be easily destroyed by static discharge. The key to protecting such circuits is to bring everything around you to a common potential. If there is no difference in charge, there is no voltage, and therefore no static buildup. For example, if you're grounded and the integrated circuits you're working with are grounded, there is no difference in charge between you and them. There is no problem. Below is a list of what you can do to keep you and the circuits you're working on at the same potential:

- Be sure all CMOS integrated circuits remain packaged in antistatic materials until they are used. Such materials include specially coated plastic carriers, conductive foam, and aluminum foil. The iea is to keep all pins of an IC shorted together and thus at the same potential. In that way no charge can build up between them.
- When handling CMOS devices, avoid touching the pins. That will keep any charge you might have accumulated off the component leads, which are, of course, connected to the solid-state material within the IC package.
- Never install or remove integrated circuits while power is still applied to the circuit. The sudden voltage jolt that could result may damage them.
- If possible, use an antistatic wrist strap. The strap brings you to ground potential. Assuming your components are at ground too, no charge can build up.

• Use a soldering iron with a grounded tip. In that way no static charge will transfer to the sensitive components while they are being soldered in the circuit.

The idea is to prevent a static buildup in the first place. If that is not possible, at least make sure that any such charge has a path to leak away. Doing that will ensure safe operation of your circuits.

Now take a moment to complete the Safety Quiz that follows.

Safety Quiz

1. List four rules for keeping the work environment safe:

   a. _____

   _____

   b. _____

   _____

   c. _____

   _____

   d. _____

   _____

2. If you are not sure how to operate a piece of test equipment, a tool, or any machinery safely, ask your _____ .

3. You should always work with one _____ behind your back while testing live circuits.

4. You should make sure all _____ are discharged before trouble-shooting begins.

5. You should always use three-conductor _____ line cords and _____ plugs with AC-operated equipment.

6. You should never install or remove any electronic components while the circuit is connected to a _____ .

7. CMOS integrated circuits can be destroyed by _____ electricity.

8. The most common way of creating a static charge is through

   _____ .

9. When handling CMOS devices, you should avoid touching the

   _____ .

10. All CMOS devices should remain packaged in _____ materials until used.

1 Basic Logic Gates

OBJECTIVES

After completing this experiment, you will be able to

- Use solderless circuit board for experimenting
- Verify AND gate logic
- Verify OR gate logic
- Verify NOT gate logic
- Verify NAND gate logic
- Verify NOR gate logic
- Determine that unused TTL inputs "float" HIGH

REFERENCE READING

Review Ronald Reis, *Digital Electronics Through Project Analysis*, Chapter 3, Section 3.2.

EQUIPMENT & MATERIALS NEEDED

Equipment

☐ 1 5-V power supply
☐ 1 logic pulser
☐ 1 logic probe
☐ 1 solderless circuit board

Materials

☐ 1 7400 quad two-input NAND gate
☐ 1 7402 quad two-input NOR gate
☐ 1 7404 hex inverter
☐ 1 7408 quad two-input AND gate
☐ 1 7432 quad two-input OR gate
☐ 2 1-kΩ resistors
☐ 1 package of jumper wires

BACKGROUND INFORMATION

In this experiment you will investigate the operation of five basic logic gates: the AND, OR, NOT, NAND, and NOR gates. In later experiments you will use these gates in many practical circuits. Here we describe them briefly.

AND Gate

An AND gate has two or more inputs and one output. Its output will be HIGH only when all inputs are HIGH. If any or all inputs are LOW, the output will be LOW. An AND gate is used to sense when inputs generating a HIGH are occurring simultaneously. It is also used to "gate through," or allow to pass, a series of pulses.

A number of ICs with individual AND gates are available to the user. Typical is the TTL 7408 quad two-input AND gate IC, the pin configuration of which is shown in Figure 1.1a. It contains four AND gates, each with two inputs. They can be used individually or together. Note that pin 7 is ground and that pin 14 is V_{CC}.

OR Gate

An OR gate has two or more inputs and one output. Its output will be HIGH when any or all inputs are HIGH. Only when all inputs are LOW will the output be LOW. The OR gate is used to output a HIGH when any input goes HIGH as a result of sensor output or signal application.

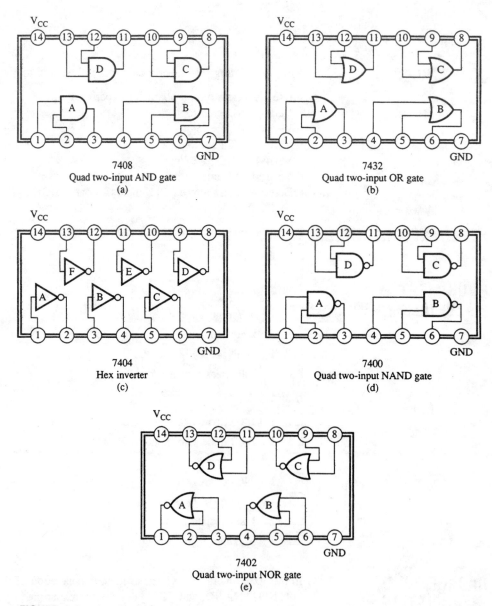

7408
Quad two-input AND gate
(a)

7432
Quad two-input OR gate
(b)

7404
Hex inverter
(c)

7400
Quad two-input NAND gate
(d)

7402
Quad two-input NOR gate
(e)

FIGURE 1.1

16

As with an AND gate, a number of ICs with individual OR gates are available to the user. Typical is the TTL 7432 quad two-input OR gate IC, the pin configuration of which is shown in Figure 1.1b. It contains four OR gates, each with two inputs. They can be used individually or together. Pin 7 is ground and pin 14 is V_{CC}.

NOT Gate

A NOT gate, also known as an inverter, has one input and one output. A NOT gate inverts the input signal: If the input is LOW, the output will be HIGH; if the input is HIGH, the output will be LOW.

A popular inverter IC is the TTL 7404 hex inverter, the pin configuration of which is shown in Figure 1.1c. It contains six NOT gates that can be used individually or together. Pin 7 is ground and pin 14 is V_{CC}.

NAND Gate

A NAND gate has two or more inputs and one output. It can be thought of as an AND gate followed by a NOT gate (the N in NAND means NOT). Thus its output is opposite to that for an AND gate. In a NAND gate, only when all inputs are HIGH will the output be LOW; if any input is LOW, the gate's output is HIGH.

The NAND gate is the basic building block in the TTL family. A typical quad two-input NAND gate is the 7400 IC, the pin configuration of which is shown in Figure 1.1d. With enough 7400s, virtually any digital circuit can be created. Pin 7 is ground and pin 14 is V_{CC}.

NOR Gate

A NOR gate has two or more inputs and one output. It can be thought of as an OR gate followed by a NOT gate. Thus its output is opposite to that for an OR gate. In a NOR gate, only when all inputs are LOW will the output be HIGH; if any input is HIGH, the gate's output is LOW.

Like the NAND gate, the NOR gate is a universal building block. Given enough NOR gates, any digital circuit can be built. The TTL 7402 IC is a typical quad two-input NOR gate. The pin configuration for the 7402 is shown in Figure 1.1e. Pin 7 is ground and pin 14 is V_{CC}.

PROCEDURE

Powering Up and Confirming the AND Gate Truth Table

☐ 1. Using solderless circuit board, install the 7408 IC as shown in Figure 1.2. Be sure the IC straddles the center channel. Standard practice is to place the IC such that pin 1 is to the lower left.

☐ 2. Apply power to the IC. To do so, connect a jumper wire from pin 7 to ground and one from pin 14 to V_{CC} (+5 V). Use your logic probe to confirm that power is getting directly to the IC pins. (*Note:* Turn off the power supply while constructing a circuit.)

FIGURE 1.2 +5v

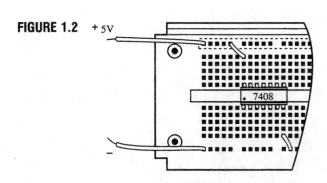

3. Using one gate of the 7408 IC (gate A), confirm the two-input AND gate truth table:
   a. Construct the circuit shown in Figure 1.3. With a jumper wire open, a given input is brought HIGH through a 1-kΩ resistor. When the jumper is closed, the input is brought LOW to ground.
   b. Set up the input conditions shown in Table 1.1 (gate A) and, using a logic probe to indicate the output, record the output logic levels in the space provided.

TABLE 1.1   7408 Quad two-input AND gate

Gate A			Gate B			Gate C			Gate D		
Input A	Input B	Output X	Input A	Input B	Output X	Input A	Input B	Output X	Input A	Input B	Output X
LO	LO		LO	LO		LO	LO		LO	LO	
LO	HI		LO	HI		LO	HI		LO	HI	
HI	LO		HI	LO		HI	LO		HI	LO	
HI	HI		HI	HI		HI	HI		HI	HI	

4. Repeat step 3 for the remaining gates (B, C, and D) in the 7408 IC. Record your results in Table 1.1 in the space provided.

Confirming the OR Gate Truth Table

1. Using solderless circuit board, install the 7432 IC and wire the power connections: ground to pin 7, +5 V to pin 14.
2. Using one OR gate of the 7432 IC (gate A), confirm the two-input OR gate truth table:
   a. Construct the circuit in Figure 1.4. Apply power.

FIGURE 1.3

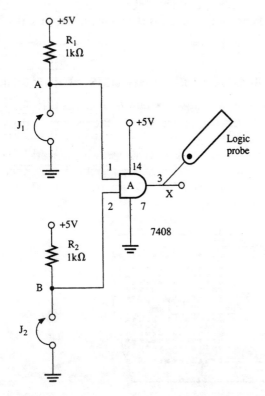

FIGURE 1.4

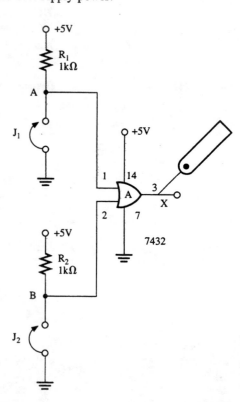

☐   b. Set up the input conditions shown in Table 1.2 (gate A) and, using a logic probe to indicate the output, record the output logic levels in the space provided.

☐   3. Repeat step 2 for the remaining gates (B, C, and D) in the 7432 IC. Record your results in Table 1.2 in the space provided.

Confirming the NOT Gate Truth Table

☐   1. Using solderless circuit board, install the 7404 IC and wire the power connections: ground to pin 7, +5 V to pin 14.

☐   2. Using one NOT gate of the 7404 IC (gate A), confirm the NOT gate truth table:

  ☐   a. Construct the circuit in Figure 1.5. Apply power.

  ☐   b. Set up the input conditions shown in Table 1.3 and, using a logic probe to indicate the output, record the output logic levels in the space provided.

☐   3. Repeat step 2 for the remaining gates (B, C, D, E, and F) in the 7404 IC. Record your results in Table 1.3 in the space provided.

TABLE 1.2 7432 Quad two-input OR gate

Gate A			Gate B			Gate C			Gate D		
Input A	Input B	Output X	Input A	Input B	Output X	Input A	Input B	Output X	Input A	Input B	Output X
LO	LO		LO	LO		LO	LO		LO	LO	
LO	HI		LO	HI		LO	HI		LO	HI	
HI	LO		HI	LO		HI	LO		HI	LO	
HI	HI		HI	HI		HI	HI		HI	HI	

TABLE 1.3 7404 Hex inverter

Gate A		Gate B		Gate C		Gate D		Gate E		Gate F	
Input A	Output X	Input A	Output X	Input A	Output X	Input A	Output X	Input A	Output X	Input A	Output X
LO		LO		LO		LO		LO		LO	
HI		HI		HI		HI		HI		HI	

FIGURE 1.5

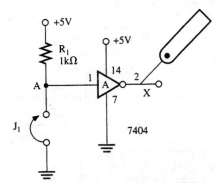

Confirming the NAND Gate Truth Table

☐ 1. Using solderless circuit board, install the 7400 IC and wire the power connections: ground to pin 7, +5 V to pin 14.

☐ 2. Using one NAND gate of the 7400 IC (gate A), confirm the two-input NAND gate truth table:

☐   a. Construct the circuit in Figure 1.6. Apply power.

☐   b. Set up the input conditions shown in Table 1.4 and, using a logic probe to indicate the output, record the output logic levels in the space provided.

☐ 3. Repeat step 2 for the remaining gates (B, C, and D) in 7400 IC. Record your results in Table 1.4 in the space provided.

Confirming the NOR Gate Truth Table

☐ 1. Using solderless circuit board, install the 7402 IC and wire the power connections: ground to pin 7, +5 V to pin 14.

☐ 2. Using one NOR gate of the 7402 (gate A), confirm the two-input NOR gate truth table:

☐   a. Construct the circuit in Figure 1.7. Apply power.

☐   b. Set up the input combinations shown in Table 1.5 and, using a logic probe to indicate the output, record the output logic levels in the space provided.

☐ 3. Repeat step 2 for the remaining gates (B, C, and D) in the 7402 IC. Record your results in Table 1.5 in the space provided.

TABLE 1.4 7400 Quad two-input NAND gate

Gate A			Gate B			Gate C			Gate D		
Input A	Input B	Output X	Input A	Input B	Output X	Input A	Input B	Output X	Input A	Input B	Output X
LO	LO		LO	LO		LO	LO		LO	LO	
LO	HI		LO	HI		LO	HI		LO	HI	
HI	LO		HI	LO		HI	LO		HI	LO	
HI	HI		HI	HI		HI	HI		HI	HI	

TABLE 1.5 7402 Quad two-input NOR gate

Gate A			Gate B			Gate C			Gate D		
Input A	Input B	Output X	Input A	Input B	Output X	Input A	Input B	Output X	Input A	Input B	Output X
LO	LO		LO	LO		LO	LO		LO	LO	
LO	HI		LO	HI		LO	HI		LO	HI	
HI	LO		HI	LO		HI	LO		HI	LO	
HI	HI		HI	HI		HI	HI		HI	HI	

FIGURE 1.6

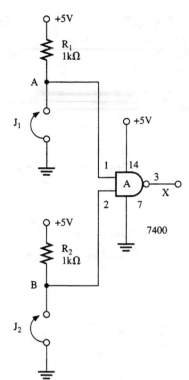

7400

FIGURE 1.7

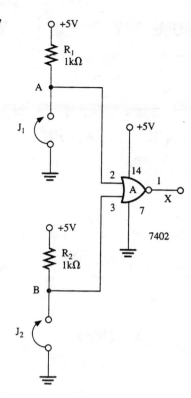

7402

Determining That Unused TTL Inputs "Float" HIGH

☐ 1. With TTL, unused inputs "float" HIGH, close to the supply voltage. They should not be left to do so, since false signals may be generated. Therefore all unused inputs must go somewhere—either tied HIGH, tied LOW, tied to other inputs, or connected to the output of a previous gate. To confirm that TTL unused inputs are HIGH:

    ☐ a. Connect the circuit shown in Figure 1.8 (gate A). Apply power.

    ☐ b. Use a logic probe to determine the logic level at the output. Record the results in Table 1.6 in the space provided.

☐ 2. Repeat the above procedure for the remaining gates (B, C, and D) in the 7408 IC. Record the results in Table 1.6 in the space provided.

FIGURE 1.8

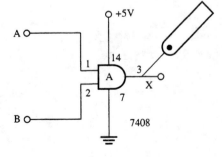

7408

TABLE 1.6

Output	
Gate A	
Gate B	
Gate C	
Gate D	

SUMMARY

In this first, "getting acquainted with logic gates," experiment, you learned to use solderless circuit board and to verify operation of the AND, OR, NOT, NAND, and NOR gates. You also determined that TTL unused inputs "float" HIGH. In the experiments to come, you will have occasion to examine these basic logic gates more fully through lots of practical applications.

REVIEW QUESTIONS

1. A(n) _____ gate is used to sense when inputs generating a HIGH are occurring simultaneously.

2. A(n) _____ gate is used to output a HIGH when any input goes HIGH as a result of sensor or signal application.

3. A(n) _____ gate inverts an input signal.

4. A(n) _____ gate is an AND gate followed by a NOT gate.

5. A(n) _____ gate is an OR gate followed by a NOT gate.

2 Exclusive Gates

OBJECTIVES

After completing this experiment, you will be able to

- Verify XOR and XNOR gate logic
- Create an even/odd parity detector
- Build and analyze a Controlled Inverter Circuit
- Build and analyze a Phase Detector Circuit

REFERENCE READING

Review Ronald Reis, *Digital Electronics Through Project Analysis*, Chapter 3, Section 3.3.

EQUIPMENT & MATERIALS NEEDED

Equipment

☐ 1 5-V power supply
☐ 1 dual-channel oscilloscope
☐ 1 logic probe
☐ 1 555 clock generator
☐ 1 solderless circuit board

Materials

☐ 1 7404 hex inverter
☐ 2 7486 quad two-input exclusive-OR gates
☐ 1 74LS266 quad two-input exclusive-NOR gate
☐ 1 LED (red)
☐ 1 LED (green)
☐ 2 220-Ω resistors
☐ 3 1-KΩ resistors
☐ 1 package of jumper wires

BACKGROUND INFORMATION

There are two exclusive gates: the exclusive-OR (XOR) and the exclusive-NOR (XNOR). The XOR gate is an *inequality* gate: When both of its inputs are unequal, the output is

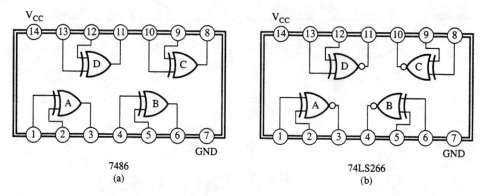

FIGURE 2.1   (a) XOR gate and (b) XNOR gate.

HIGH. The XOR gate is an *equality* gate: When both of its inputs are the same, the output is HIGH.

Exclusive gates are used in comparator, parity detector, phase detector, controlled inverter, and binary adder circuits—to name but a few. You will examine a number of these applications shortly.

As with basic logic gates, XOR and XNOR gates are available in integrated-circuit packages. Typical of the former is the 7486 TTL quad two-input exclusive-OR gate. An example of the latter would be the TTL 74LS266 quad two-input exclusive-NOR gate (open collector). The pin configurations for both ICs are shown in Figure 2.1.

In this experiment you will use the 7486 and 74LS266 ICs to investigate exclusive-gate operation.

PROCEDURE

Part 1: Circuit Fundamentals

Confirming XOR and XNOR Truth Tables

- [ ] 1. Using solderless circuit board, install the 7486 XOR IC and wire the power connections: ground to pin 7, +5 V to pin 14.
- [ ] 2. Using one XOR gate of the 7486 (gate A), confirm the two-input XOR gate truth table:
  - [ ] a. Construct the circuit in Figure 2.2. Apply power.
  - [ ] b. Set up the input combinations shown in Table 2.1 and, using a logic probe to indicate the outputs, record the output logic levels in the space provided.
- [ ] 3. Repeat step 2 for the remaining gates (B, C, and D) in the 7486 IC. Record your results in Tables 2.2, 2.3, and 2.4 in the space provided.
- [ ] 4. Using solderless circuit board, install the 74LS266 XNOR IC and wire the power connections: ground to pin 7, +5 V to pin 14.

TABLE 2.1   Gate A

A	B	Record output X
LO	LO	
LO	HI	
HI	LO	
HI	HI	

TABLE 2.2   Gate B

A	B	Record output X
LO	LO	
LO	HI	
HI	LO	
HI	HI	

TABLE 2.3   Gate C

A	B	Record output X
LO	LO	
LO	HI	
HI	LO	
HI	HI	

TABLE 2.4   Gate D

A	B	Record output X
LO	LO	
LO	HI	
HI	LO	
HI	HI	

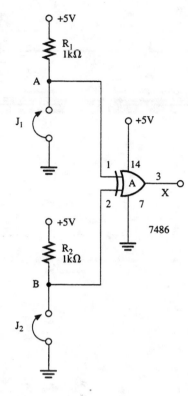

FIGURE 2.2

7486

5. Using one XNOR gate of the 74LS266 (gate A), confirm the two-input XNOR gate truth table:

  a. Construct the circuit in Figure 2.3. Apply power. Note the 1-kΩ pull-up resistor on the gate's output. Remember, this is an open collector chip; thus its output must be pulled HIGH.

  b. Set up the input combinations shown in Table 2.5 and, using a logic probe to indicate the outputs, record the output logic levels in the space provided.

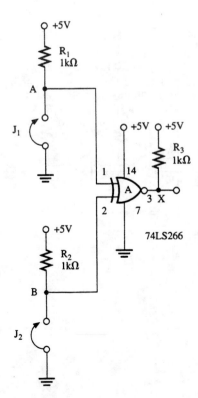

FIGURE 2.3

74LS266

TABLE 2.5   Gate A

A	B	Record output X
LO	LO	
LO	HI	
HI	LO	
HI	HI	

25

Part 2: Further Investigation

Even and Odd Parity Checking

☐ 1. *Parity* refers to the oddness or evenness of the number of 1s in a specified group of bits. A circuit using a combination of XOR gates can check for even parity (an even number of 1s). With the addition of one NOT gate, the circuit can also indicate odd parity (an odd number of 1s). A circuit that will do both is shown in Figure 2.4.

    ☐ a. Using solderless circuit board, construct the circuit of Figure 2.4. Apply power.

    ☐ b. Check for even parity by tying an even number of inputs HIGH, directly to +5 V, and the rest LOW to ground. Which LED is lit?_____

       Why is this so? _____

       _____

       Trace through the circuit with a pencil to confirm the logic.

    ☐ c. Check for odd parity by tying an odd number of inputs HIGH, directly to +5 V, and the rest LOW to ground. Which LED is lit? _____

       Why is this so? _____

       _____

       Trace through the circuit with a pencil to confirm the logic.

Part 3: Circuit Applications

The Controlled Inverter: Controlled Inverter Circuit

☐ 1. A simple controlled inverter circuit can be built using just one XOR gate, as shown in Figure 2.5. With one input (pin 2) as the control, the signal on the other input (pin 1) can be made to pass through uninverted or inverted. With pin 2

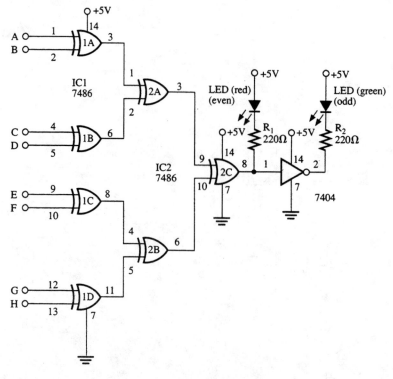

FIGURE 2.4

26

FIGURE 2.5

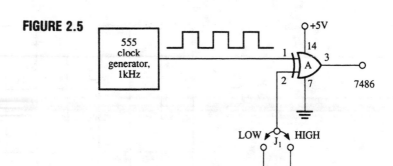

LOW, the input signal passes uninverted. With pin 2 HIGH, the input signal is inverted at the output.

☐ a. Using solderless circuit board, construct the circuit of Figure 2.5. Apply power. With the 555 clock generator set at 1 kHz, inject a signal into pin 1.

☐ b. Also connect one channel of the dual-channel oscilloscope to pin 1. Obtain a stable wave pattern on the screen.

Connect the second channel of the oscilloscope to pin 3.

☐ c. Tie pin 2 LOW to ground. Is the output signal appearing on channel 2 of the oscilloscope inverted or uninverted with reference to the input signal?

_____ Why is this so? _____

_____

Draw the input and output waveforms in Plot 2.1.

☐ d. Now tie pin 2 HIGH to +5 V. Is the output signal (appearing on channel 2 of the oscilloscope) inverted or uninverted with reference to the input

signal? _____ Why is this so? _____

_____

Draw the input and output waveform in Plot 2.2.

Phase Detection: Phase Detector Circuit

☐ 1. A single XOR gate coupled to a NOT gate can be used to form a simple Phase Detector Circuit, as shown in Figure 2.6. When the two input signals are out of phase, the LED glows. When they are in phase, it does not.

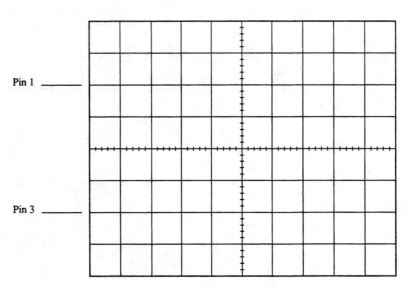

Pin 1 _____

Pin 3 _____

PLOT 2.1

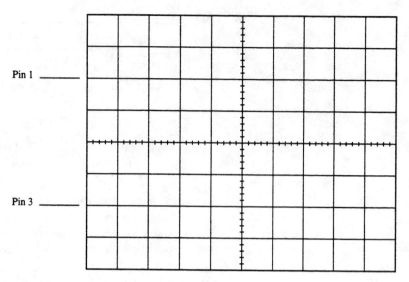

PLOT 2.2

FIGURE 2.6

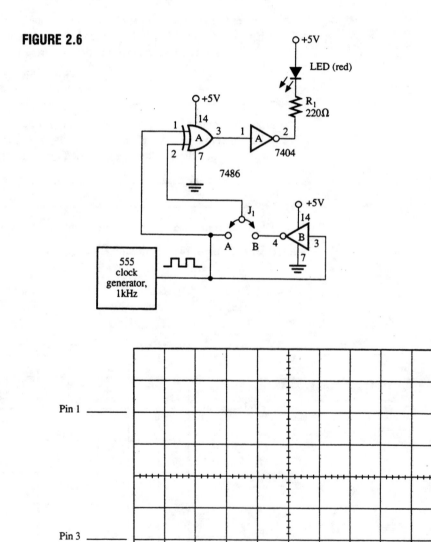

PLOT 2.3

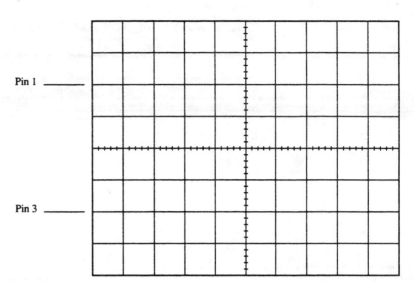

Pin 1 _____

Pin 3 _____

PLOT 2.4

☐ a. Using solderless circuit board, construct the circuit of Figure 2.6. Apply power. Set the 555 clock generator to 1 kHz. Place jumper J_1 in position A so that the 1-kHz square wave is sent to both inputs of the XOR gate. Using a dual-channel oscilloscope, monitor both input signals. Is the LED lit? _____ Why is this so? _____

Draw the two input waveforms in Plot 2.3.

☐ b. Now place jumper J_1 in position B. What happens to the input signal on pin 2 in reference to the signal on pin 1? _____ Why is this so? _____

Is the LED lit? _____ Why is this so? _____

Draw the two input waveforms in Plot 2.4.

SUMMARY

In this experiment you verified XOR and XNOR logic, assembled and examined an even/odd parity checker, and built and analyzed two application circuits: the controlled inverter and the phase detector.

REVIEW QUESTIONS

1. Draw the schematic symbols for a two-input XOR gate and a two-input XNOR gate.

2. Is it possible for XOR and XNOR gates to have more than two inputs? _____ Explain. _____

_____

3. Draw the schematic for a three-input XOR gate using two two-input XOR gates.

4. Using two two-input XOR gates, design a two-input XNOR gate.

Writing Skills Assignment

In as few paragraphs as possible, explain how either the Controlled Inverter Circuit or the Phase Inverter Circuit works. You might first discuss what the circuit does and then discuss how it does it.

3 Gating Principles

OBJECTIVES

After completing this experiment, you will be able to

- Verify operation of a gating circuit that will block or pass an input signal.
- Verify operation of a gating circuit that will pass an input signal, inverted or noninverted
- Build and analyze the operation of a Gated Audio Oscillator Circuit

REFERENCE READING

Review Ronald Reis, *Digital Electronics Through Project Analysis*, Chapter 5, Section 5.1.

EQUIPMENT & MATERIALS NEEDED

Equipment

- [ ] 1 5-V power supply
- [ ] 1 555 clock generator
- [ ] 1 dual-channel oscilloscope
- [ ] 1 solderless circuit board

Materials

- [ ] 1 555 timer IC
- [ ] 1 7402 quad two-input NOR gate
- [ ] 1 7486 quad two-input exclusive-OR (XOR) gate
- [ ] 1 2N3904 NPN transistor
- [ ] 1 4.7-μF capacitors
- [ ] 1 100-Ω resistor
- [ ] 2 1-kΩ resistors
- [ ] 1 1-MΩ resistor
- [ ] 1 8-Ω speaker
- [ ] 1 N.O. push-button switch
- [ ] 1 package of jumper wires

BACKGROUND INFORMATION

Gating is based on the principle that a voltage applied to one input of a logic gate (the control input) can be used to control a signal arriving at the other input (the signal input). Gating is of two types. In the first case (Figure 3.1a), the input signal being controlled, or gated, will either be blocked or else be allowed to pass to the output, contingent on whether the control input voltage level is LOW or HIGH. If the signal passes, it will appear either inverted or noninverted at the output, depending on the type of gate used. In the second case (Figure 3.1b), the signal always passes, but depending on the voltage level on the control input, the signal is either inverted or noninverted at the output.

In this experiment you will use a two-input OR gate to test the first type of gating circuit and a two-input exclusive-OR (XOR) gate to check the second type. You will also build and analyze a Gated Audio Oscillator Circuit and thus see how the gating principle is actually applied.

PROCEDURE

Part 1: Circuit Fundamentals

Blocking or Passing a Signal

☐ 1. With a two-input OR gate, if the control input is HIGH, the input signal cannot pass to the output because the output is always HIGH. If the control input is LOW, however, the input signal will pass to the output, noninverted. The two-input OR gate circuit of Figure 3.2 illustrates the concept. (Note that to keep the selection of chips in this first part to a minimum, we use two two-input NOR gates from a TTL 7402 IC to create one two-input OR gate. Gate B, of course, is simply acting as an inverter.)

    ☐ a. Construct the circuit of Figure 3.2. Apply power.

    ☐ b. Using the 555 clock generator set to 1 kHz, place a signal on pin 3 of the 7402 IC.

    ☐ c. Use channel 1 of your dual-channel oscilloscope to display the signal input. Use channel 2 to display the signal output.

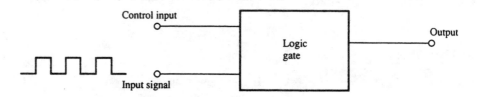

• Control input determines whether input signal is blocked or allowed to pass.

• Type of gate determines whether the signal passed is inverted or non-inverted.

FIRST CASE
(a)

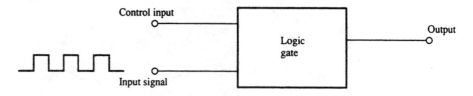

• Control gate determines whether input signal is inverted or non-inverted at the output. Exclusive gates are used.

SECOND CASE
(b)

FIGURE 3.1   (a) First type of gating and (b) second type of gating.

FIGURE 3.2

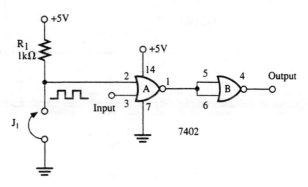

☐   d. Open jumper J_1. Describe the output signal. _____

_____

       Explain why it appears this way. _____

_____

☐   e. Close J_1. Describe the output signal. _____

_____

☐   f. Draw and label the input and output signal waveforms in Plot 3.1.

☐   2. With a two-input NOR gate, if the control input is HIGH, the input signal cannot pass to the output because the output is always LOW. If the control input is LOW, however, the input signal will pass to the output, inverted. The two-input NOR gate circuit of Figure 3.3 illustrates the concept.

    ☐   a. Construct the circuit of Figure 3.3. Apply power.

    ☐   b. Using the 555 clock generator set to 1 kHz, place a signal on pin 3 of the 7402 IC.

    ☐   c. Use channel 1 of your dual-channel oscilloscope to display the signal input. Use channel 2 to display the signal output.

    ☐   d. Open jumper J_1. Describe the output signal. _____

       Why is this so? _____

_____

    ☐   e. Close J_1. Describe the output signal. _____

    ☐   f. Draw and label the input and output signal waveforms in Plot 3.2.

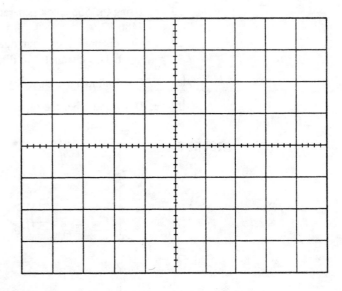

PLOT 3.1

FIGURE 3.3

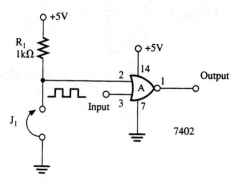

PLOT 3.2

Part 2: Further Investigation

Inverted or Noninverted Output

☐ 1. With a two-input XOR gate, the control input determines whether the input signal is inverted or noninverted at the output. A test circuit, using a TTL 7486 quad two-input exclusive OR gate IC, is shown in Figure 3.4.

    ☐ a. Construct the circuit of Figure 3.4. Apply power.

    ☐ b. Using the 555 clock generator set to 1 kHz, place a signal on pin 2 of the 7486 IC.

    ☐ c. Use channel 1 of your dual-channel oscilloscope to display the signal input. Use channel 2 to display the signal output.

    ☐ d. Open jumper J_1. Describe the output signal. _____

    ☐ e. Draw and label the input and output signal waveforms in Plot 3.3.

FIGURE 3.4

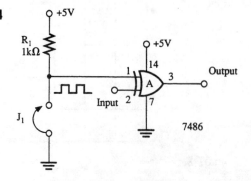

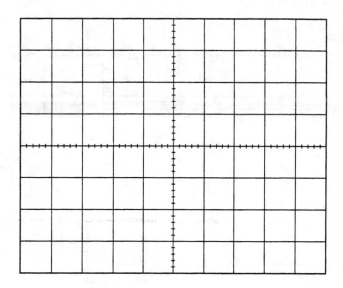

PLOT 3.3

☐   f. Close J_1. Describe the output signal. _____

☐   g. Draw and label the input and output signal waveforms in Plot 3.4.

☐   h. Summarize your conclusions with regard to Part 2. _____

_____

_____

Part 3: Circuit Applications

Applying the Gating Principle: Gated Audio Oscillator Circuit

☐   1. The Gated Audio Oscillator Circuit of Figure 3.5 illustrates clearly the gating principle. When switch S_1 of the 555 one-shot circuit is momentarily pressed, the output at pin 3 goes HIGH for a period of time determined by the values of R_2 and C_1. (With the values shown, the output stays HIGH for approximately 5 s.) When pin 3 is HIGH, the output of inverter gate B is LOW. Thus the LOW on control input pin 2 of NOR gate A allows the 1-kHz signal from the 555 clock

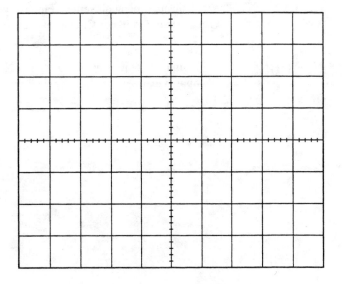

PLOT 3.4

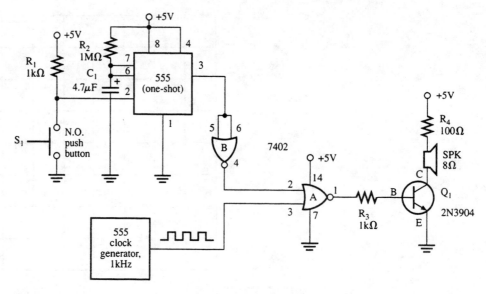

FIGURE 3.5

generator to pass, inverted, to transistor Q_1. The speaker puts out a 1-kHz tone. When the timing cycle is complete, the output of the 555 one-shot circuit goes LOW and the output of gate B goes HIGH. The HIGH on control pin 2 of gate A now prevents the signal produced by the 555 clock generator from reaching the NOR gate's output. The speaker is mute.

☐ a. Construct the circuit of Figure 3.5. Apply power.

☐ b. With switch S_1 left open, is there any sound from the speaker?

_____ Explain why there is or is not a sound. _____

_____

☐ c. Momentarily press S_1. Is there any sound from the speaker?

_____ If so, for approximately how long is it heard?

SUMMARY

In this experiment you began by assembling and testing a two-input OR gate circuit that would block or pass, noninverted, an input signal. Then, using a two-input NOR gate, you saw how a similar circuit would block an input signal or pass it to the output, inverted. With a two-input XOR gate, you used the control input to pass an input signal, either inverted or noninverted. Finally, you built and then analyzed the operation of a Gated Audio Oscillator Circuit that illustrates the basic gating principle.

REVIEW QUESTIONS

1. Design a circuit using a two-input AND gate that will block or pass, noninverted, an input signal.

2. Design a circuit using a two-input NAND gate that will block or pass, inverted, an input signal.

3. If the gate of Figure 3.4 is replaced with a two-input XNOR gate, and if J_1 is open, what will the output signal be? _____

4. What is the purpose of gate A in Figure 3.5? _____

_____

Writing Skills Assignment

In as few paragraphs as possible, explain how the Gated Audio Oscillator Circuit works. You might first discuss what the circuit does and then discuss how it does it.

4

Schmitt Triggers

OBJECTIVES

After completing this experiment, you will be able to

• Verify operation of a Schmitt trigger as a wave shaper
• Verify operation of a Schmitt trigger as a threshold detector
• Build and analyze a Bounceless Switch and an LED Flasher Circuit using a single Schmitt trigger gate

REFERENCE READING

Review Ronald Reis, *Digital Electronics Through Project Analysis*, Chapter 5, Section 5.1.

EQUIPMENT & MATERIALS NEEDED

Equipment

☐ 1 5-V power supply
☐ 1 dual-channel oscilloscope
☐ 1 function generator
☐ 1 solderless circuit board

Materials

☐ 1 74LS132 quad two-input NAND Schmitt trigger
☐ 1 LED (red)
☐ 1 220-Ω resistor
☐ 1 1-kΩ resistor
☐ 1 47-μF capacitor
☐ 1 100-μF capacitor
☐ 1 N.O. push-button switch
☐ 1 package of jumper wires

BACKGROUND INFORMATION

A Schmitt trigger is a logic gate that produces a sharp, crisp square wave regardless of whether the input waveform has a long rise or fall time, a large noise element, or varying amplitude pulses. The Schmitt trigger works by having an upper and lower trip point, referred to as the UTP and LTP, respectively. As shown with the NOT gate of Figure 4.1,

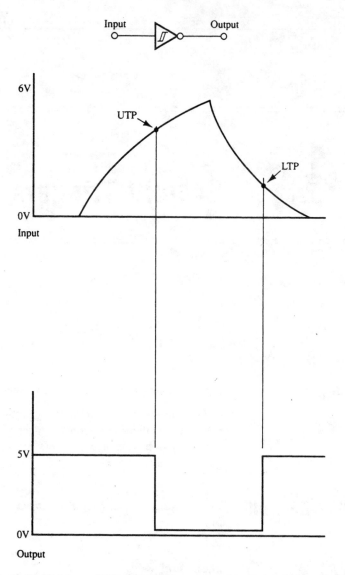

FIGURE 4.1   Schmitt trigger NOT gate (UTP = upper trip point; LTP = lower trip point).

when the input *exceeds* the UTP, the output "snaps" to a LOW state. (Remember, this is an inverter.) It remains there until the input drops *below* the LTP; then the output "snaps" to a HIGH state. (Note that the UTP and LTP are not at the same voltage level.) As a result of this upper- and lower-trip-point characteristic, the Schmitt trigger refuses to respond to every "wiggle" or slight variation in input signal. It only changes states when an input voltage level exceeds the UTP or drops below the LTP.

A Schmitt trigger is usually set up to perform *wave shaping, pulse restoration, noise elimination,* or *threshold detection.* In all cases, the basic idea is the same: The output changes states only when the input voltage level crosses above the UTP or below the LTP.

In this experiment you will use the TTL 74LS132 quad two-input NAND Schmitt trigger to investigate Schmitt-trigger operation. First you will see how the Schmitt trigger, wired as a NOT gate, shapes incoming sine and sawtooth waves. In both cases, you will identify the upper and lower trip points. Then you will examine the concept of threshold detection by varying the amplitude of an incoming square wave to see the effect at the Schmitt trigger's output. Finally, to more fully appreciate the Schmitt trigger's toggle (or "snap-action") feature, you will build and analyze two single-gate Schmitt-trigger circuits, the Bounceless Switch and the LED Flasher.

PROCEDURE

Part 1: Circuit Fundamentals

Wave Shaping

☐ 1. To examine the Schmitt trigger as a wave shaper, you will begin by seeing how the circuit reacts to a sine-wave input.

    ☐ a. Construct the circuit in Figure 4.2a. Apply power.

    ☐ b. Using your function generator, input a sine wave of 5 V peak to peak at approximately 1.5 kHz.

    ☐ c. Use channel 1 of your dual-channel oscilloscope to display the sine-wave input and channel 2 to display the square-wave output. Superimpose the inverted square wave over the sine wave, as shown in Figure 4.2b.

    ☐ d. Examine the waveform carefully. Note that the UTP and LTP are not at the same voltage level. At what voltage point on the rising curve of the sine wave (UTP) does the output "snap" to a LOW voltage?

    _____ At what voltage point on the falling curve of the sine

    wave (LTP) does the output "snap" to a HIGH voltage? _____

    Are the upper and lower trip points the same? _____

☐ 2. If your function generator can produce a sawtooth wave, examine how the Schmitt trigger inverter deals with it.

    ☐ a. Construct the circuit of Figure 4.3a. Apply power.

    ☐ b. Using your function generator, input a sawtooth wave of 5 V peak to peak at approximately 1.5 kHz.

    ☐ c. Use channel 1 of your dual-channel oscilloscope to display the sawtooth input and channel 2 to display the square-wave output. Superimpose the inverted square wave over the sawtooth wave as shown in Figure 4.3b.

    ☐ d. At what voltage point on the rising line of the sawtooth wave (UTP) does

    the output "snap" to a LOW voltage? _____ At what voltage point on the falling line of the sawtooth wave (LTP) does the output

    "snap" to a HIGH voltage? _____ Are the upper and lower trip

    points the same? _____

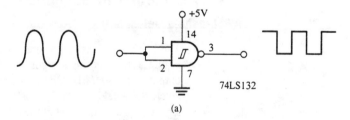

(a)

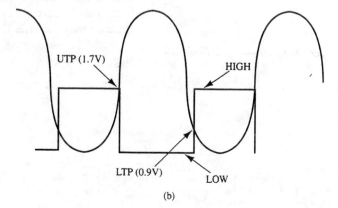

(b)

FIGURE 4.2

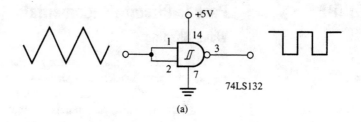

(a)

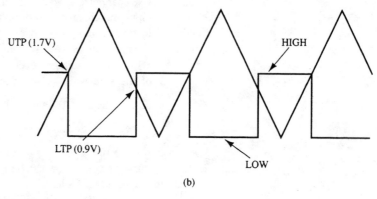

UTP (1.7V)

HIGH

LTP (0.9V)

LOW

(b)

FIGURE 4.3

☐ 3. In reference to steps 1 and 2, what are your conclusions with regard to how a Schmitt trigger handles non-square-wave inputs? _____

_____

_____

_____

Part 2: Further Investigation

Threshold Detection

☐ 1. The Schmitt trigger can also act as an excellent threshold detector. As shown in Figure 4.4, if the input signal level is below the threshold level, the inverter's output remains HIGH. On the other hand, if the input signal exceeds the threshold level, the gate's output "snaps" LOW. Note, also, that even though the input signal exceeds the threshold level, the output remains at a fixed LOW voltage.

   ☐ a. Construct the circuit shown in Figure 4.4. Apply power.

   ☐ b. Using your function generator, input a square wave with a peak of 2 V at approximately 1.5 kHz. Use channel 1 of your dual-channel oscilloscope to display the incoming square wave and channel 2 to display the outgoing inverted square wave.

   ☐ c. Is there an output square wave? _____

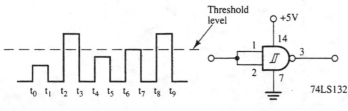

Threshold level

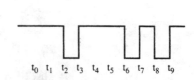

FIGURE 4.4

42

    ☐   d. Increase the amplitude of the incoming square wave until an outgoing square wave appears. At what threshold level does this occur?

    _____

    ☐   e. Continue to increase the amplitude of the incoming square wave. What happens to the outgoing square wave? _____ Why is this so? _____

    _____

    ☐   f. Summarize your conclusions with regard to Part 2. _____

    _____

    _____

Part 3: Circuit Applications

The "Snap-Action" Effect: Bounceless Switch and LED Flasher Circuit

☐ 1. A simple but effective bounceless switch can be made using one NAND gate Schmitt trigger wired as an inverter, a resistor, a capacitor, and a normally open push-button switch. The circuit is shown in Figure 4.5. With S_1 open, C_1 will have charged to the supply voltage through R_1. The gate's input is HIGH and its output LOW. When S_1 is momentarily closed, C_1 discharges through S_1 and the gate's input is brought LOW. Its output, of course, "snaps" HIGH, with a clean, bounceless action.

    ☐   a. Construct the circuit in Figure 4.5. Apply power.

    ☐   b. Use channel 1 of your dual-channel oscilloscope to monitor the inverter's output.

    ☐   c. Press and hold S_1 closed. What happens to the output signal? _____

    Why is this so? _____

    _____

    ☐   d. Momentarily press S_1. What happens to the output signal? _____

    Why is this so? _____

    _____

☐ 2. A simple LED flasher can be built using one Schmitt-trigger gate, two resistors, a capacitor, and an LED, as shown in Figure 4.6. Assume, initially, that the gate's input is LOW. Its output therefore is HIGH; the LED is on, and capacitor C_1 begins to charge through R_1. When the voltage across C_1 reaches the upper trip point (1.7 V for a 74LS132), the gate's output rapidly switches to a LOW. The LED turns off. The capacitor now discharges through R_1 and through the LOW output of the gate. When the voltage on the capacitor falls below the lower

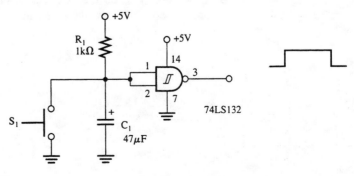

FIGURE 4.5

43

FIGURE 4.6

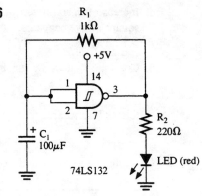

trip point (0.9 V for a 74LS132), the gate's output swings back to a HIGH. The process then repeats.

☐ a. Construct the circuit in Figure 4.6. Apply power.

☐ b. Does the LED flash? _____ At approximately what rate?

_____

SUMMARY

In this experiment you began by examining the Schmitt trigger as a wave-shaping circuit. You then explored how it acts as a threshold detector. You concluded by building and analyzing two circuits, each using only one Schmitt-trigger inverter, a Bounceless Switch and an LED Flasher Circuit.

REVIEW QUESTIONS

1. Draw the schematic for a two-input NAND gate Schmitt trigger.

2. What is the basic difference between a standard inverter gate and a Schmitt-trigger inverter gate? _____

_____

_____

_____

3. How would you convert the circuit of Figure 4.6 to a "controlled LED flasher" with an enable input? Draw the schematic.

4. How would you decrease the LED flash rate of the circuit shown in Figure 4.6? ___

_____

_____

Writing Skills Assignment

In as few paragraphs as possible, explain how either the Bounceless Switch Circuit or the LED Flasher Circuit works. You might first discuss what the circuit does and then discuss how it does it.

5 Magnitude Comparators

OBJECTIVES

After completing this experiment, you will be able to

- Assemble a magnitude comparator circuit
- Verify operation of a 4-bit magnitude comparator
- Assemble and verify operation of an 8-bit magnitude comparator
- Build and analyze a Magnitude Comparator HI-LO Game Circuit

REFERENCE READING

Review Ronald Reis, *Digital Electronics Through Project Analysis*, Chapter 5, Section 5.1.

EQUIPMENT & MATERIALS NEEDED

Equipment

- ☐ 1 5-V power supply
- ☐ 1 555 clock generator
- ☐ 1 solderless circuit board

Materials

- ☐ 2 7485 4-bit magnitude comparators
- ☐ 1 7493 binary counter
- ☐ 1 LED (red)
- ☐ 1 LED (yellow)
- ☐ 1 LED (green)
- ☐ 1 220-Ω resistor
- ☐ 16 1-kΩ resistors
- ☐ 1 N.O. push-button switch
- ☐ 2 8-position DIP switches (or equivalent)
- ☐ 1 package of jumper wires

BACKGROUND INFORMATION

A magnitude comparator compares two input words and indicates which is larger or if they are equal. A block diagram of a 4-bit comparator is shown in Figure 5.1. It compares two 4-bit words, P and Q.

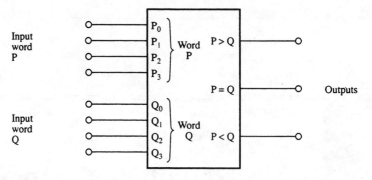

FIGURE 5.1

For example, if word P is equal to word Q, output P = Q goes HIGH while outputs P > Q and P < Q go LOW. If word P is greater than word Q, output P > Q is HIGH; outputs P =Q and P < Q are LOW. And if word Q is greater than word P, output P < Q will be HIGH while outputs P = Q and P > Q stay LOW.

A typical 4-bit magnitude comparator is the TTL 7485, the pin configuration of which is shown in Figure 5.2. It has four input lines for word P (P_0, P_1, P_2, and P_3); four input lines for word Q (Q_0, Q_1, Q_2, and Q_3); three output lines (P > Q, P = Q, and P < Q); and three expansion inputs (>, =, <), which are used when cascading comparators. Pin 8 is ground; pin 16 is V_{CC}.

In this experiment you will use the 7485 IC to investigate magnitude comparators.

PROCEDURE

Part 1: Circuit Fundamentals

Confirming Magnitude Comparator Operation

☐ 1. Referring to Figure 5.3, record missing IC pin numbers. Using solderless circuit board, construct the circuit shown in Figure 5.3. Apply power. Confirm magnitude comparator operation:

    ☐ a. Input the nibble 0111 for word P, via DIP switches S_0–S_3. Input the nibble 0111 for word Q, via DIP switches S_4–S_7. The P = Q "red LED" should light. The green and yellow LEDs should be off. Does this

        happen? _____ Why is this so? _____

        _____

    ☐ b. Leave word Q at 0111. Input a nibble greater than 0111 for word P.

        Which LED is lit? _____ Why is this so? _____

        _____

FIGURE 5.2

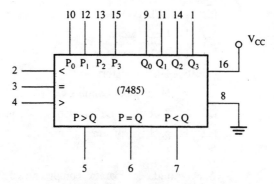

7485 4-bit magnitude comparator

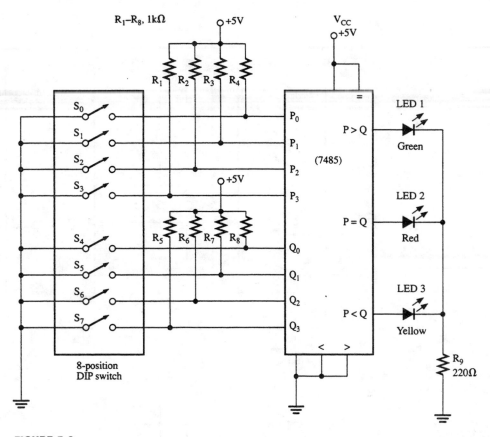

FIGURE 5.3

☐   c. Once more leaving word Q at 0111, input a nibble less than 0111 for word P. Which LED is lit? _____ Why is this so? _____

_____

_____

☐   2. Set up additional nibble combinations for words P and Q. Again confirm magnitude comparator operation. Summarize your results:

_____

_____

_____

Part 2: Further Investigation

Creating an 8-Bit Magnitude Comparator

☐   1. Two 4-bit 7485s can be cascaded to produce an 8-bit magnitude comparator, as shown in Figure 5.4. Note that the outputs of the least significant comparator, IC1, are fed to the expansion inputs of the most significant comparator, IC2. Note also the switch positions for the 8-bit words. For word P, bits 0–3 (the word's least significant nibble) are on switches S_0–S_3 of IC1. For bits 4–7 (the word's most significant nibble), switches S_0–S_3 of IC2 are used. The same pattern is true for the 8-bit Q word. Bits 0–3 (the word's least significant nibble) are on switches S_4–S_7 of IC1. For bits 4–7 (the word's most significant nibble), switches S_4–S_7 of IC2 are used.

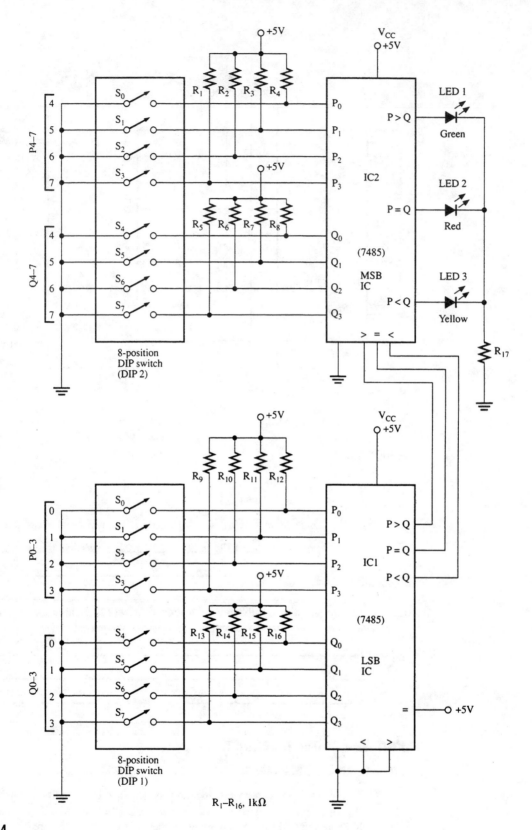

FIGURE 5.4

Referring to Figure 5.4, record missing IC pin numbers. Using your solderless circuit board, construct the circuit of Figure 5.4. Apply power. Confirm 8-bit magnitude comparator operation:

☐   a. Input the byte 01110011 for word Q. Input the same byte for word P.

Which LED is lit? _____ Why is this so? _____

_____

_____

☐   b. Leave word Q with the byte 01110011. Input the byte 01110100 for word P. Which LED is lit? _____ Why is this so? _____

_____

_____

☐   c. Once more leaving word Q at 01110011, input the byte 01110010 for word P. Which LED is lit? _____ Why is this so? _____

_____

_____

☐   2. Set up additional byte combinations for words P and Q. Again confirm magnitude comparator operation. Summarize your results.

_____

_____

_____

Part 3: Circuit Applications

Checking Out a Magnitude Comparator the Fun Way: Magnitude Comparator HI-LO Game Circuit

☐   1. To better understand the 7485 IC, why not have a little fun with it? Let's build the Magnitude Comparator HI-LO Game Circuit shown in Figure 5.5. The circuit consists of a TTL 7493 binary counter, IC1, that counts from 0000_2 to 1111_2 and recycles as clock pulses arrive from the clock generator via push-button switch S_4. The 7493's outputs (Q_0–Q_3) are sent to the P_0–P_3 inputs of the 7485, IC2. The Q_0–Q_3 inputs of the magnitude comparator receive their signals from DIP switches S_0–S_3.

     The idea of the game is to guess the bit pattern of the nibble arriving from IC1. If you guess right, the red LED will light. If the incoming nibble is higher than your guess, the green LED will turn on. If the incoming nibble is lower than your guess, the yellow LED will glow.

     Referring to Figure 5.5, record missing IC pin numbers. Using solderless circuit board, construct the circuit of Figure 5.5. Apply power.

☐   a. Set the 555 clock generator to approximately 1 kHz.

☐   b. Next, set the DIP switches to your "guess," that is, place a bit pattern (nibble) on inputs Q_0–Q_3 somewhere between 0000_2 and 1111_2.

☐   c. Press and hold S_4 to allow the 1-kHz signal to repeatedly cycle the 7493 through its count. Do all the LEDs flash or appear to be on?

_____ Why is this so? _____

_____

_____

Release S_4. Which LED remains on? _____ What does this indicate? _____

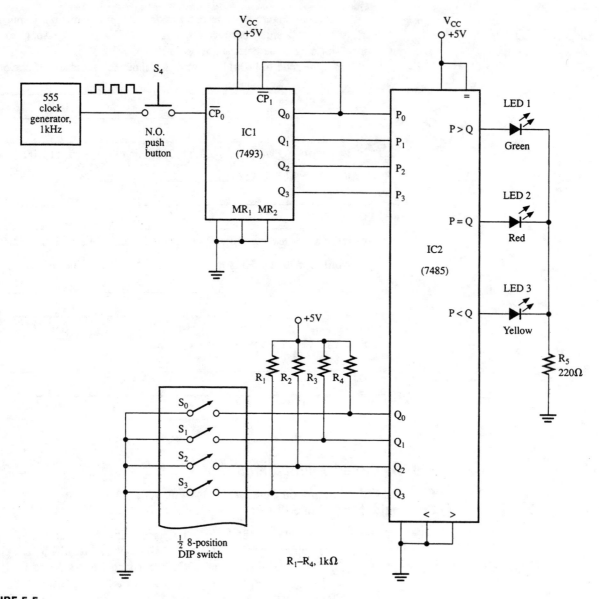

FIGURE 5.5

☐ 2. Set up additional guesses and play the game some more. Summarize your results:

_____

_____

_____

SUMMARY

In this experiment you assembled and verified operation of a 4-bit magnitude comparator. You also cascaded two such comparators to form an 8-bit magnitude comparator. You then had a little fun by building and analyzing a Magnitude Comparator HI-LO Game Circuit.

REVIEW QUESTIONS

1. A magnitude comparator compares two input words and indicates which is _____ or if they are the same.
2. In the circuit of Figure 5.3, if the A word is 0110 and the B word is 1011, which LED will light? _____

3. Why does the circuit of Figure 5.3 use only one current-limiting resistor, R$_9$, instead of three (one for each LED)? _____

_____

4. Draw a block diagram for a 12-bit magnitude comparator that uses three 7485 ICs.

Writing Skills Assignment

In as few paragraphs as possible, explain how the Magnitude Comparator HI-LO Game Circuit works. You might first discuss what the circuit does and then discuss how it does it.

6 Encoders

OBJECTIVES

After completing this experiment, you will be able to

- Assemble and verify operation of a discrete encoder
- Assemble and verify operation of an 8-line-to-3-line MSI priority encoder
- Build and analyze an Octal Keypad Scanner Circuit

REFERENCE READING

Review Ronald Reis, *Digital Electronics Through Project Analysis*, Chapter 5, Section 5.1, and Chapter 7, Section 7.1.

EQUIPMENT & MATERIALS NEEDED

Equipment

- ☐ 1 5-V power supply
- ☐ 1 555 clock generator
- ☐ 1 solderless circuit board

Materials

- ☐ 1 7400 quad two-input NAND gate
- ☐ 1 7493 binary counter
- ☐ 1 74148 priority encoder
- ☐ 1 74151 data selector/multiplexer
- ☐ 1 LED (red)
- ☐ 1 LED (yellow)
- ☐ 1 LED (green)
- ☐ 3 220-Ω resistors
- ☐ 8 1-kΩ resistors
- ☐ 1 8-position DIP switch (or equivalent)
- ☐ 1 package of jumper wires

BACKGROUND INFORMATION

An encoder converts a single input, such as a key press representing a number, letter, or graphic symbol, into an equivalent binary-coded output. In this experiment you will concentrate on number encoders, that is, those that convert an octal, decimal, or

FIGURE 6.1

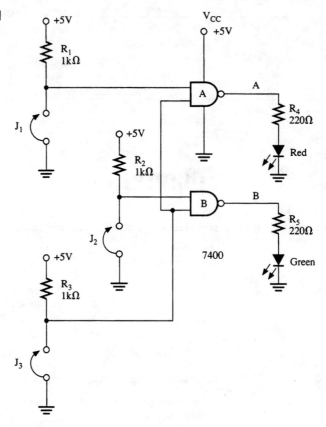

hexadecimal number into an identical binary number. First you will assemble and test a discrete two-gate encoder that converts the decimal numbers 1 to 3 into binary equivalents. Then you will explore the MSI 74148 priority encoder, examining how it works and how it acts to prioritize input numbers. Finally, you will build and analyze an Octal Keypad Scanner Circuit that uses the 74151 multiplexer chip.

PROCEDURE

Part 1: Circuit Fundamentals

A Discrete Encoder

☐ 1. To analyze basic encoding, you can build a simple discrete decimal 1–3-to-binary encoder using just two two-input NAND gates, as shown in Figure 6.1. This is an active-LOW encoder; when a given data input, 1–3, goes LOW, a binary equivalent output goes HIGH, turning on an appropriate LED. The circuit's truth table is shown in Table 6.1. Note that when data input number 1

TABLE 6.1

Input			Binary Outputs		Outputs measured	
J_1	J_2	J_3	(2) B	(1) A	(2) B	(1) A
1	1	0	1	1		
1	0	1	1	0		
0	1	1	0	1		

is LOW, output A is HIGH and output B is LOW. When data input number 2 is LOW, output A is LOW and output B is HIGH. And when input 3 is LOW, both outputs A and B are HIGH.

☐   a. Referring to Figure 6.1, record missing IC pin numbers. Construct the circuit in Figure 6.1. Apply power.

☐   b. Place each successive data-input LOW while keeping the other inputs HIGH. Record the output logic levels in the appropriate space in Table 6.1. Do the LEDs light in proper sequence, indicating the correct binary

output? _____

Part 2: Further Investigation

An MSI Encoder

☐ 1. You don't have to build encoders using discrete gates; MSI chips exist to do the job. An example is the TTL 74148 (octal) priority encoder. This IC will encode eight data lines to three-line (4−2−1) binary (octal). It also ensures that only the highest-order data line is encoded. The pin configuration diagram for the 74148 is shown in Appendix A (Figure A.20). The truth table is shown in Table 6.2.

☐ 2. A test circuit for the 74148 is shown in Figure 6.2. Note that enable input $\bar{E}_1$ is tied LOW, while outputs $\bar{G}_S$ and $\bar{E}_0$ are placed LOW and HIGH, respectively. Furthermore, observe that all data inputs and outputs are active-LOW.

☐   a. Referring to Figure 6.2, record missing IC pin numbers. Construct the circuit of Figure 6.2. Apply power.

☐   b. Bring data-in line I_0 LOW while keeping all others HIGH. Do any LEDs

light? _____ Why is this so? _____

_____

☐   c. Next, place each successive data-in line (I_1–I_7) LOW while keeping all other inputs HIGH. Do the LEDs light in proper sequence, indicating the

Inputs									Outputs				
$\bar{E}_1$	I_0	I_1	I_2	I_3	I_4	I_5	I_6	I_7	(C) A_2	(B) A_1	(A) A_0	$\bar{G}_s$	$\bar{E}_0$
1	x	x	x	x	x	x	x	x	1	1	1	1	1
0	1	1	1	1	1	1	1	1	1	1	1	1	0
0	x	x	x	x	x	x	x	0	0	0	0	0	1
0	x	x	x	x	x	x	0	1	0	0	1	0	1
0	x	x	x	x	x	0	1	1	0	1	0	0	1
0	x	x	x	x	0	1	1	1	0	1	1	0	1
0	x	x	x	0	1	1	1	1	1	0	0	0	1
0	x	x	0	1	1	1	1	1	1	0	1	0	1
0	x	0	1	1	1	1	1	1	1	1	0	0	1
0	x	1	1	1	1	1	1	1	1	1	1	0	1

TABLE 6.2

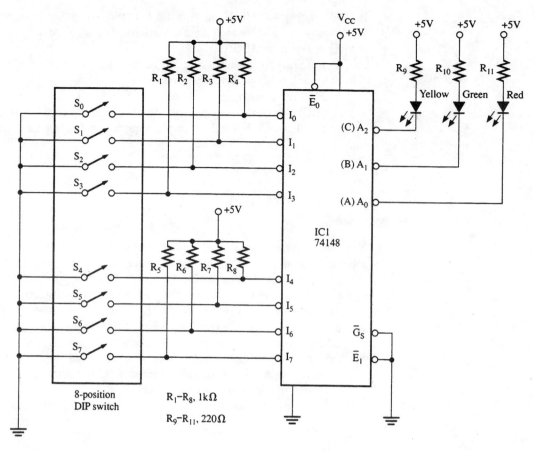

FIGURE 6.2

correct binary (001 – 111) output? _____ (Remember, an active-LOW output will turn on a corresponding LED.)

☐ 3. The 74148 is also a priority encoder, meaning it will encode only the highest-order data-in line. To test, place an input, say number I_3, LOW. Which

LEDs come on? _____ Now, keeping input I_3 LOW, place

a higher input LOW, say number I_6. Which LEDs are lit? _____

Why is this so? _____

Part 3: Circuit Applications

Keypad Scanning: Octal Keypad Scanning Circuit

☐ 1. To encode more than 10 lines, a method known as *keypad scanning* is used. The circuit of Figure 6.3 demonstrates the process. Note first that it has only eight input lines. To keep circuit complexity to a minimum, but still illustrate the procedure, we have chosen to present an 8- rather than a 16-input keypad scanner. The circuit consists of a single two-input NAND gate, a 7493 binary counter, and a 74151 multiplexer. Here is how it works:

With the Y output of the 74151 HIGH, the 1-kHz pulse train from the clock generator is passed through the NAND gate to arrive on the $\overline{CP}_0$ pin of the 7493. As a result, IC2 counts up rapidly in binary on outputs Q_0, Q_1, and Q_2, and then recycles. Since the select lines of IC3 are connected directly to the outputs of IC2 (Q_0, Q_1, and Q_3), the 74151 scans its data-in lines, I_0–I_7, in equally rapid succession, over and over again.

When a given input line (I_0–I_7) is brought LOW, by the closing of an appropriate DIP switch, output Y of IC3 immediately goes LOW. As a result, the

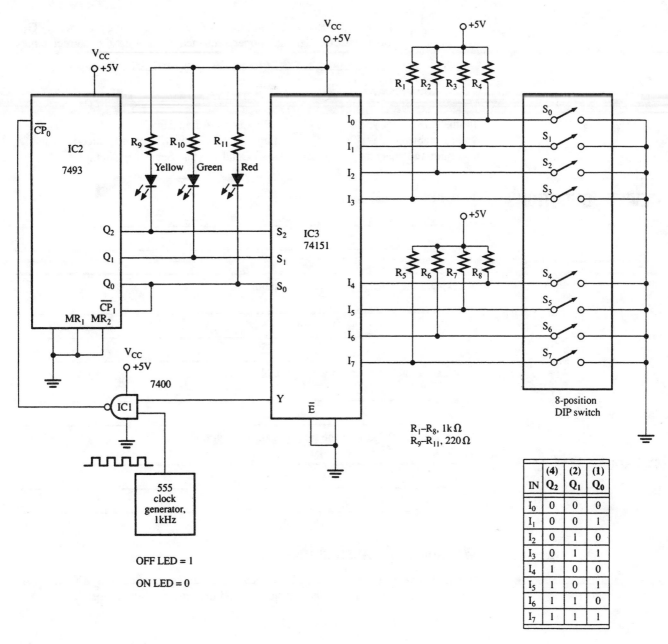

FIGURE 6.3

NAND gate's output "locks" on HIGH, pulses are prevented from reaching IC2, and the count on lines Q_0, Q_1, and Q_2 "freezes." The address LEDs identify, in binary, the data-in line that was brought LOW.

One more thing: As the chart in Figure 6.3 indicates, an off LED represents a 1, an on LED represents a 0. This is the reverse of what might be expected.

☐ a. Referring to Figure 6.3, record missing IC pin numbers. Construct the circuit in Figure 6.3. Apply power.

☐ b. Open all DIP switches, thus bringing all data-in lines HIGH. Are the three address LEDs lit? _____ Why is this so? _____

_____

☐ c. Bring data-in line I_0 LOW while keeping all others HIGH. What is the status of the three LEDs: Red _____? Green _____? Yellow _____?

☐ d. Place each successive data-in line LOW while keeping all others HIGH. Do the LEDs light in proper sequence, indicating the correct binary output? _____ (Remember, an off LED indicates a 1, an on LED, a 0.)

SUMMARY

In this experiment you began by assembling and testing a simple discrete encoder. Next you assembled and verified operation of an 8-line-to-3-line MSI priority encoder. Finally, you built and analyzed the operation of an Octal Keypad Scanning Circuit.

REVIEW QUESTIONS

1. Write a definition for an encoder.

   _____
   _____
   _____

2. Explain the concept of priority encoding.

   _____
   _____
   _____

3. What do the Xs mean in the truth table of Table 6.2?

   _____
   _____

4. Explain the purpose of the NAND gate in Figure 6.3.

   _____
   _____
   _____

Writing Skills Assignment

In as few paragraphs as possible, explain how the Octal Keypad Scanning Circuit works. You might first discuss what the circuit does and then discuss how it does it.

7 Decoders

OBJECTIVES

After completing this experiment, you will be able to

- Assemble and verify operation of a discrete decoder
- Assemble and verify operation of an MSI BCD-to-decimal decoder
- Build and analyze a 10-LED Chaser Circuit

REFERENCE READING

Review Ronald Reis, *Digital Electronics Through Project Analysis,* Chapter 5, Section 5.1 and Chapter 7, Section 7.1.

EQUIPMENT & MATERIALS NEEDED

Equipment

☐ 1 5-V power supply
☐ 1 logic probe
☐ 1 555 clock generator
☐ 1 solderless circuit board

Materials

☐ 1 7400 quad two-input NAND gate
☐ 1 7404 hex inverter
☐ 1 7442 BCD-to-decimal decoder
☐ 1 7490 decade counter
☐ 10 LEDs (red)
☐ 4 220-Ω resistors
☐ 4 1-kΩ resistors
☐ 1 8-position DIP switch (or equivalent)
☐ 1 package of jumper wires

BACKGROUND INFORMATION

A decoder changes binary code into a recognizable number, letter, or symbol. A number decoder is the opposite of a number encoder. Instead of identifying 1-of-8, -10, or -16 inputs and creating a binary-weighted output, as an encoder does, a decoder selects 1-of-8, -10, or -16 outputs, depending of the ABCD input code.

In this experiment you will concentrate on number encoders. First you will assemble and test a 1-of-4 decoder using four NOT gates and four two-input NAND gates. Next you will explore the 7442 BCD-to-decimal decoder, examining how it decodes 1-of-10 (0–9) outputs and handles invalid (10–15) input conditions. Finally, you will build and analyze a 10-LED Chaser Circuit that has many exciting and useful applications.

PROCEDURE

Part 1: Circuit Fundamentals

A 1-of-4 Discrete Decoder

☐ 1. To analyze basic decoding, a simple 1-of-4 decoder using four NOT and four NAND gates can be built, as shown in Figure 7.1. As you know, it takes two binary input lines to decode four output lines. With LOWs on inputs A and B, output 0 goes LOW and LED D_0 lights. All other outputs are HIGH, and thus all other LEDs are off. In a similar manner, the 01, 10, and 11 binary inputs select the 1, 2, and 3 output lines, respectively. Using your pencil, trace through the 0s and 1s of the circuit to convince yourself that the correct decoding is indeed taking place.

   ☐ a. Referring to Figure 7.1, record missing IC pin numbers. Construct the circuit of Figure 7.1. Apply power.

   ☐ b. By opening or closing jumpers J_1 and J_2, work your way through Table 7.1, completing the truth table as you go. Remember, an on LED indicates a LOW output.

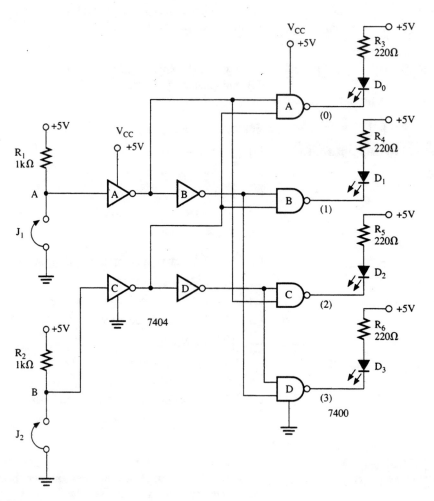

FIGURE 7.1

TABLE 7.1

B	A	0	1	2	3
0	0				
0	1				
1	0				
1	1				

☐ c. Does only one LED light for each binary input? _____ Do the LEDs "advance" in sequence, as the input count goes from 00 to 11 in binary? _____

Part 2: Further Investigation

An MSI BCD-to-Decimal Decoder

☐ 1. As with encoders, you do not have to build discrete decoders; many MSI versions exist. One such decoder is the TTL 7442 BCD-to-decimal decoder, the pin configuration for which is shown in Appendix A (Figure A.6). In addition to ground and V_{CC}, it has 10 (0–9) active-LOW outputs that are selected by four BCD (A_0, A_1, A_2, A_3) inputs. Since it is a decimal, not a hexadecimal, decoder, input conditions 10–15 are considered *invalid*. If any input combination 1010 through 1111 is selected on the A_0, A_1, A_2, A_3 inputs, *all* outputs go HIGH.

☐ 2. A test circuit for the 7442 is shown in Figure 7.2. A partially completed truth table is shown in Table 7.2.

☐ a. Referring to Figure 7.2, record missing IC pin numbers. Construct the circuit of Figure 7.2. Apply power.

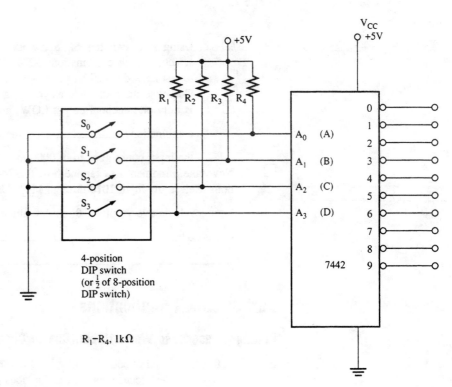

FIGURE 7.2

No.	(8) A_3	(4) A_2	(2) A_1	(1) A_0	0	1	2	3	4	5	6	7	8	9
	BCD input				**Decimal output**									
0	0	0	0	0										
1	0	0	0	1										
2	0	0	1	0										
3	0	0	1	1										
4	0	1	0	0										
5	0	1	0	1										
6	0	1	1	0										
7	0	1	1	1										
8	1	0	0	0										
9	1	0	0	1										
I n v a l i d	1	0	1	0	1	1	1	1	1	1	1	1	1	1
	1	0	1	1	1	1	1	1	1	1	1	1	1	1
	1	1	0	0	1	1	1	1	1	1	1	1	1	1
	1	1	0	1	1	1	1	1	1	1	1	1	1	1
	1	1	1	0	1	1	1	1	1	1	1	1	1	1
	1	1	1	1	1	1	1	1	1	1	1	1	1	1

TABLE 7.2

☐ b. Using DIP switches S_0–S_3, work your way through the truth table, selecting input combinations 0000–1001. As you do, use a logic probe to test *each* decimal output (0–9) for *every* input combination selected. Complete the truth table as you go along.

☐ c. Is only one decimal output LOW, and all others HIGH, for each input combination? _____ Do the LOWs "advance" in sequence, as the input count goes from 0000 to 1001? _____

☐ 3. Now take a moment to examine how the 7442 handles invalid input combinations. Select any input 1010 through 1111. Using your logic probe, measure the logic levels on every output, 0–9. Are any outputs LOW? _____ Why is this so? _____

_____

Part 3: Circuit Applications

Putting a Decoder to Work: 10-LED Chaser Circuit

☐ 1. A fun and useful attention-getting LED chaser circuit can be built using a 7442 decoder, a 7490 decade counter, and a clock generator, as shown in Figure 7.3. The circuit will flash the LEDs sequentially at the clock input frequency. Such a

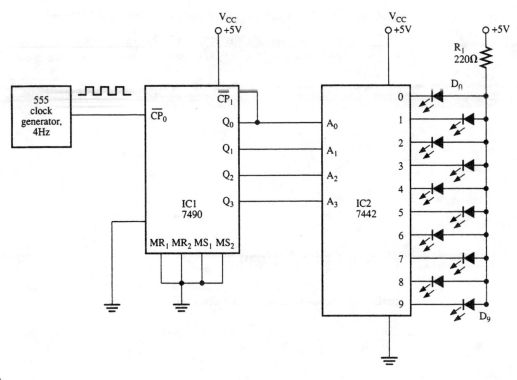

FIGURE 7.3

device is ideal for special lighting effects, costumes, and so forth. Operation is as follows:

Clock pulses arriving on the $\overline{CP}_0$ input of IC1 cause the decade counter to count up in binary from 0000 to 1001 and then recycle. Since the select input lines of IC2 are connected directly to the BCD outputs of IC1, each output of the 7442 is brought LOW in sequence, lighting a respective LED.

☐ a. Referring to Figure 7.3, record missing IC pin numbers. Construct the circuit of Figure 7.3. Apply power.

☐ b. With the 555 clock generator set to 4 Hz, the LEDs should sequence through in 2½ s. Do they? _____

☐ c. Increase the 555 clock generator frequency to 100 Hz. What happens to the LEDs? _____ Why is this so? _____

_____

SUMMARY

In this experiment you began by assembling and testing a simple 1-of-4 discrete decoder. Next you assembled and verified operation of an MSI BCD-to-decimal decoder. Finally, you built and analyzed the operation of a 10-LED Chaser Circuit.

REVIEW QUESTIONS

1. Write a definition for a decoder.

_____

_____

2. Explain why the 7442 IC has six invalid inputs.

_____

_____

_____

3. What two TTL ICs can you use to replace the 7490 and 7442, respectively, of Figure 7.3 in order to convert the circuit to a 16-LED chaser circuit? _____

4. Draw the schematic for a 16-LED chaser circuit using the two ICs selected in Review Question 3.

Writing Skills Assignment

In as few paragraphs as possible, explain how the 10-LED Chaser Circuit works. You might first discuss what the circuit does and then discuss how it does it.

8 Seven-Segment Displays and Decoders

OBJECTIVES

After completing this experiment, you will be able to

- Test a seven-segment common-anode LED display
- Test a seven-segment liquid-crystal display
- Verify operation of a TTL 7447 BCD-to-seven-segment decoder/driver
- Build and analyze a Single-Digit Seven-Segment Display Demonstration Circuit

REFERENCE READING

Review Ronald Reis, *Digital Electronics Through Project Analysis*, Chapter 7, Section 7.2.

EQUIPMENT & MATERIALS NEEDED

Equipment

☐ 1 5-V power supply
☐ 1 555 clock generator
☐ 1 solderless circuit board

Materials

☐ 1 7447 BCD-to-seven-segment decoder/driver
☐ 1 7486 quad two-input exclusive-OR gate
☐ 1 seven-segment common-anode LED display
☐ 1 seven-segment liquid-crystal display
☐ 7 220-Ω resistors
☐ 7 1-kΩ resistors
☐ 1 8-position DIP switch (or equivalent)
☐ 1 package of jumper wires

BACKGROUND INFORMATION

The ubiquitous seven-segment digital display uses seven segments to configure the numerals 0 through 9. The segments are always arranged and designated as shown in Figure 8.1. As seen in the figure, a decimal point (DP) is often included too.

The two most popular types of seven-segment displays are the light-emitting diode (LED) and liquid-crystal display (LCD). The former are of two kinds: common anode and

FIGURE 8.1

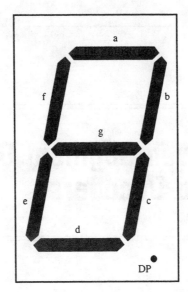

common cathode. With the common-anode LED display, all individual LED segments have their anodes tied together (in common). The common-cathode display, as you would expect, ties the cathodes of all individual LED segments together. Both kinds are shown in Figure 8.2.

LCDs also use seven segments to display the numerals 0 through 9; but instead of individual LEDs, they use liquid crystals to form each segment. In operation, low-frequency voltages are applied to the crystal's backplane and individual segments. If the voltages are in phase, the segment is transparent to the eye. If they are 180 degrees out of phase, the segment appears black. By "darkening" various segments, the numerals 0 through 9 are formed.

Both types of displays, LED and LCD, require a decoder IC to convert BCD data into a format for producing decimal digits. A number of TTL and CMOS decoder chips are available to do the job.

In this experiment you will begin by "checking out" a common-anode LED display, that is, by identifying various segments and making sure they are all working. Next you will do the same for a liquid-crystal display. You will then examine the TTL 7447 BCD-to-seven-segment decoder/driver IC, using it to drive your common-anode LED display. Finally, you will build and analyze the operation of a Single-Digit Seven-Segment Display Demonstration Circuit.

FIGURE 8.2

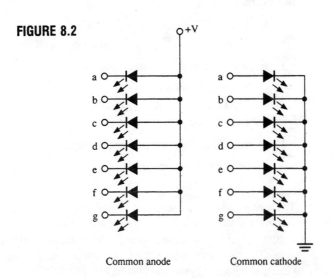

Common anode   Common cathode

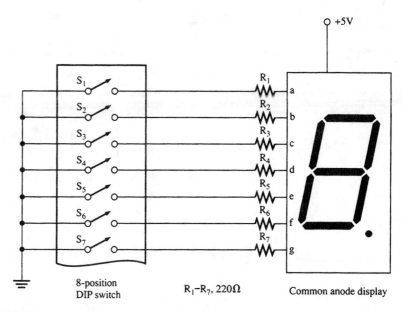

FIGURE 8.3

PROCEDURE

Part 1: Circuit Fundamentals

Testing Seven-Segment Displays

☐ 1. To test a common-anode LED display, construct the circuit shown in Figure 8.3. Using the data sheet for your particular display, identify the anode pin (there are usually two) and connect it directly to +5 V. Identify all other pins and label the figure accordingly. Connect a 220-Ω resistor from each segment pin, a through g, to one end of an SPST DIP switch, as shown. Place the correct pin numbers directly on the drawing.

    ☐ a. Close switch S_1, grounding the LED cathode of segment a, while leaving all other switches open. Does the segment light? _____

    ☐ b. Proceed to close each switch, S_1–S_7, in turn while leaving all others open. Record the segment that lights for each switch closure:

        S_1 _____; S_2 _____; S_3 _____; S_4 _____; S_5 _____;

        S_6 _____; S_7 _____

    ☐ c. Configure the numeral 5 by closing the appropriate switches. Which switches are closed? _____, _____, _____, _____, _____

    ☐ d. Configure the numeral 7 by closing the appropriate switches. Which switches are closed? _____, _____, _____

☐ 2. If you have a liquid-crystal display, test it with the circuit of Figure 8.4. This circuit supplies two 60-Hz square wave signals of opposite polarity. One is applied to the LCD's backplane, the other to a segment that you wish to appear dark.

    ☐ a. Referring to Figure 8.4, record missing IC pin numbers. Construct the circuit of Figure 8.4. Apply power.

    ☐ b. Connect a clip lead from point A directly to the backplane pin of your LCD. (*Note:* The backplane pin on most LCDs is in the lower-left corner.) Connect a clip lead from point B to a segment pin. Does a given segment go dark? _____

    ☐ c. Try moving the clip lead attached to point B to other segment pins. As you do, do various segments darken? _____

FIGURE 8.4

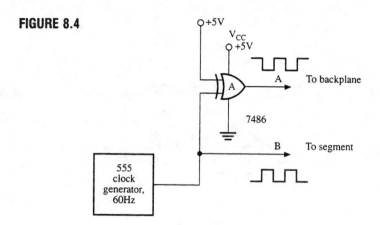

Part 2: Further Investigation

The Seven-Segment Decoder

☐ 1. The TTL 7447, the pin configuration of which is shown in Appendix A (Figure A.7) is a BCD-to-seven-segment decoder/driver featuring active-LOW outputs designed to drive a common-anode LED display. In addition, the 16-pin IC has full ripple-blanking input/output controls and a lamp-test input. A test circuit is shown in Figure 8.5.

    ☐ a. Referring to Figure 8.5, record missing IC pin numbers. Construct the circuit shown in Figure 8.5. Apply power.

    ☐ b. Open switches S_5–S_7 and close switches S_1–S_4. Does the display light?

_____ If so, what numeral appears? _____

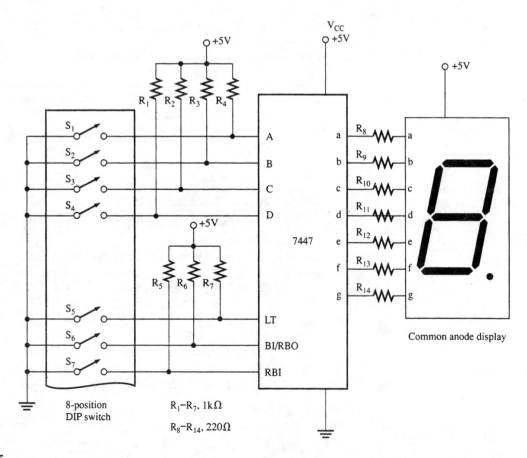

FIGURE 8.5

c. Proceed with the input BCD count (binary 0001–1001) by opening appropriate switches S_1–S_4. (The 7447 has active-HIGH BCD inputs.) In each case, record the numeral presented on the display:

0001 _____ ; 0010 _____ ; 0011 _____ ; 0100 _____ ;

0101 _____ ; 0110 _____ ; 0111 _____ ; 1000 _____ ;

1001 _____

2. In addition to providing seven-segment display decoding, the 7447 features a lamp test (LT), a leading-edge-zero blanking control (BI/RBO), and a trailing-edge-zero blanking control, RBI.

a. The lamp test will turn on all LED segments. Open switches S_1–S_4 (switches S_5–S_7 should still be open), bringing all BCD inputs, A through D, HIGH and turning off the display. Now bring active-LOW lamp test (LT) LOW by closing S_5. What happens to the display? ____

_____

Open all switches, S_1–S_7.

b. The BI/RBO input will blank the display. It is used to suppress a leading-edge zero(s) in a string of displays. Close switches S_1–S_4 while leaving switches S_5–S_7 open. What numeral appears on the display?

_____ Bring active-LOW input, BI/RBO, LOW by closing S_6.

What happens to the display? _____ Open all switches, S_1–S_7.

c. The RBI input will also blank the display, but only if it is currently indicating a zero. It is used to suppress a trailing-edge zero(s) in a string of displays. Open switches S_5–S_7. Using switches S_1–S_4, place a zero on the display. Bring the active-LOW input, RBI, LOW by closing S_7. What happens to the display? _____

Open all switches, S_1–S_7. Using switches S_1–S_4, place a numeral other than zero on the display. What is that numeral? _____ Close S_7.

What happens to the display? _____

Why is this so? _____

_____

Part 3: Circuit Applications

A Display Counter: Single-Digit Seven-Segment Display Demonstration Circuit

1. A circuit used to automatically display the digits 0 through 9, in sequence, is shown in Figure 8.6. The 7490 is a decade counter that will place active-HIGH logic levels on its outputs Q_0–Q_3 (0000–1001) as pulses arrive at its $\overline{CP_0}$ clock input. If the 555 clock generator is set to 1 Hz, the display will count up from 0 to 9 and repeat at a 1-Hz rate.

a. Referring to Figure 8.6, record missing IC pin numbers. Construct the circuit of Figure 8.6. Apply power.

b. Does the display proceed through the count 0–9 and then repeat?

_____

c. Set the 555 clock generator to 1 kHz. What does the display indicate?

_____ Why is this so? _____

_____

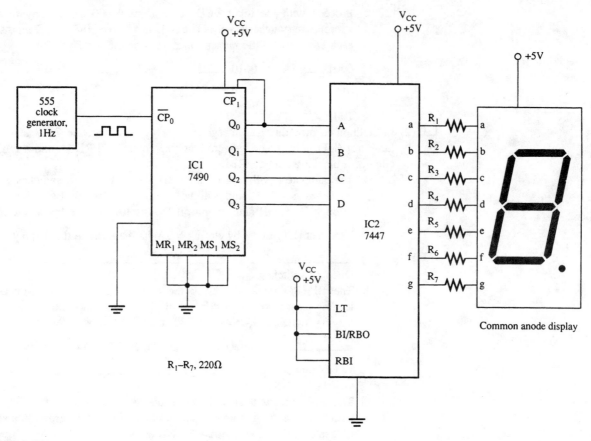

FIGURE 8.6

SUMMARY

In this experiment you began by testing a seven-segment common-anode LED display and a seven-segment liquid-crystal display. You then examined the operation of a 7447 BCD-to-seven-segment decoder/driver chip. You concluded by building and analyzing the performance of a Single-Digit Seven-Segment Display Demonstration Circuit.

REVIEW QUESTIONS

1. Explain the difference between a seven-segment common-anode and a seven-segment common-cathode LED display.

_____

_____

_____

2. Out-of-phase voltages between the backplane of an LCD and a given segment cause the segment to appear _____.

3. The 7447 IC outputs are active-_____, while the BCD inputs are active-_____.

4. The LT, BI/RBO, and RBI inputs to a 7447 IC are active-_____.

Writing Skills Assignment

In as few paragraphs as possible, explain how the Single-Digit Seven-Segment Display Demonstration Circuit works. You might first discuss what the circuit does and then discuss how it does it.

9 Multiplexers

OBJECTIVES

After completing this experiment, you will be able to

- Assemble and then verify operation of a discrete multiplexer
- Verify operation of a dual 4-line-to-1-line MSI multiplexer
- Build and analyze a Watchdog Multiplexer Circuit

REFERENCE READING

Review Ronald Reis, *Digital Electronics Through Project Analysis,* Chapter 5, Section 5.1, and Chapter 8, Section 8.1.

EQUIPMENT & MATERIALS NEEDED

Equipment

☐   1   5-V power supply
☐   1   logic probe
☐   1   logic pulser
☐   1   555 clock generator
☐   1   solderless circuit board

Materials

☐   1   7404 hex inverter
☐   1   7408 quad two-input AND gate
☐   1   7432 quad two-input OR gate
☐   1   7493 binary counter
☐   1   74153 dual 4-line-to-1-line multiplexer
☐   1   SCR C106B1
☐   4   LEDs (red)
☐   4   220-Ω resistors
☐   6   1-kΩ resistors
☐   1   8-position DIP switch (or equivalent)
☐   1   N.C. push-button switch
☐   1   package of jumper wires

A multiplexer selects one of several input lines and applies any data on that line to a single output. Hence it is also referred to as a *data selector*. There are 2-, 4-, 8-, and 16-line multiplexers available in MSI DIP packages, both TTL and CMOS. In this experiment you will begin by building a discrete two-input MUX to get a feel for how data-line selection takes place. Next you will test the TTL 74153, a dual 4-line-to-1-line multiplexer. Finally, you will use the 74153 as the basis for a Watchdog Multiplexer Circuit that illustrates the "input polling" aspect of many multiplexer circuits.

Part 1: Circuit Fundamentals

A Discrete Multiplexer

☐ 1. A simple, discrete 2-line-to-1-line multiplexer using two NOT gates, two AND gates, and one OR gate is shown in Figure 9.1. As the accompanying truth table indicates, when select input (address) line S is LOW, data on line A will be channeled to the output, Y. If the data on line A is LOW, a LOW appears at Y. If data on line A is HIGH, a HIGH appears on Y. It doesn't matter (X) what is happening on line B because it has not been selected by the select input S. (Reason: The lower input line of AND gate A is LOW; thus the gate is disabled and its output is always LOW.)

When line S is placed HIGH, the reverse takes place; now data on line B is channeled to the Y output. Any data on line A is blocked from reaching the output because pin 5 of AND gate B is LOW. The gate, then, is disabled, its output locked on LOW.

☐   a. Referring to Figure 9.1, record missing IC pin numbers. Construct the circuit in Figure 9.1. Apply power.

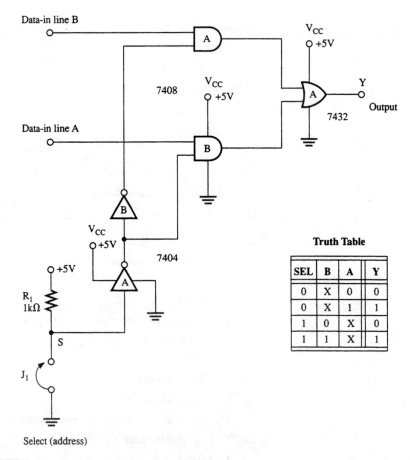

SEL	B	A	Y
0	X	0	0
0	X	1	1
1	0	X	0
1	1	X	1

FIGURE 9.1

TABLE 9.1

Step	Select	Data in B	Data in A	Output Y
1	0	x	‾ ‾ ‾	
2	0	⎍⎍⎍	0	
3	0	⎍⎍⎍	H	
4	1	⎍⎍⎍	x	
5	1	∠	⎍⎍	
6	1	1	⎍⎍	

☐ b. Place select line S LOW, as indicated in step 1 of Table 9.1. Using a logic pulser, or the 555 clock generator set to 5 Hz, place a signal on line A. It doesn't matter what you do with line B. Using a logic probe, measure the logic level at Y. Record the result—a LOW, a HIGH, or a pulse—in the space provided.

☐ c. Proceed through steps 2–6 in a similar manner, recording your results as you go. In every step, does the output receive the data from the input line

selected? _____ Why is this so? _____

_____

Part 2: Further Investigation

An MSI Multiplexer

☐ 1. To gain experience with a multiplexer chip, we have selected the TTL 74153, the pin configuration for which is shown in Appendix A (Figure A.22). There are two four-input multiplexers in each IC, designated I_a and I_b. Each multiplexer has its own active-LOW enable pin, $\overline{E}_a$ for MUX I_a and $\overline{E}_b$ for MUX I_b, as well as its own group of four data-in lines: I_{0a}, I_{1a}, I_{2a}, and I_{3a} for MUX I_a; I_{0b}, I_{1b}, I_{2b}, and I_{3b} for MUX I_b. The output for MUX I_a is Y_a; the output for MUX I_b, Y_b. Both multiplexers, however, share the same select, or address, lines, S_0 and S_1. Ground is on pin 8; V_{CC}, at pin 16.

☐ 2. A test circuit for the 74153 is shown in Figure 9.2. Which of the two multiplexers

is being used for checkout? _____

   ☐ a. Referring to Figure 9.2, record missing IC pin numbers. Construct the circuit shown in Figure 9.2. Apply power.

   ☐ b. Close jumpers J_1 and J_2 and the four DIP switches, S_0–S_3. With $\overline{E}_a$ LOW, the chip is enabled. But when $\overline{E}_a$ is HIGH, the chip is disabled. To test, tie $\overline{E}_a$ HIGH. Now place a logic probe on output Y_a. What logic

   level is indicated? _____

   ☐ c. Use J_1 and J_2 to select data-in lines I_{0a}–I_{3a} in turn. As you do, open and close the respective DIP switch, S_0–S_3. Does the logic level at output Y_a

   change? _____ Thus you can conclude that with $\overline{E}_a$ HIGH, the I_a

   multiplexer is _____; its output will always be _____.

☐ 3. Return $\overline{E}_a$ to ground. Multiplexer I_a is now enabled. To check chip operation, set up the input conditions shown in Table 9.2 and, using a logic probe, measure and record the output logic levels in the space provided.

☐ 4. Remove switches S_0–S_3 and pull-up resistors R_1–R_4 from the circuit. Using J_1 and J_2, select data-in line I_{0a}. With a logic pulser, or the 555 clock generator set

Select inputs		Input					Output
S_0	S_1	$\overline{E}_a$	I_{0a}	I_{1a}	I_{2a}	I_{3a}	Y_a
0	0	0	0	x	x	x	
0	0	0	1	x	x	x	
1	0	0	x	0	x	x	
1	0	0	x	1	x	x	
0	1	0	x	x	0	x	
0	1	0	x	x	1	x	
1	1	0	x	x	x	0	
1	1	0	x	x	x	1	

TABLE 9.2

to 5 Hz, place a signal on data-in line I_{0a}. Use your logic probe to check the output at Y_a. Are the pulses getting through? _____ Without changing J_1 and J_2, place the input signal on another data-in line. Now measure the output with a logic probe. What is the logic level? _____ Why is this so? ___

_____

☐ 5. Using J_1 and J_2, select data-in lines I_{1a}–I_{3a} in turn. As you do, apply a 5-Hz signal to the respective input. Does the signal appear at the output, Y_a? _____ What happens at the output when the 5-Hz signal is applied to a nonselected input? _____

input? _____

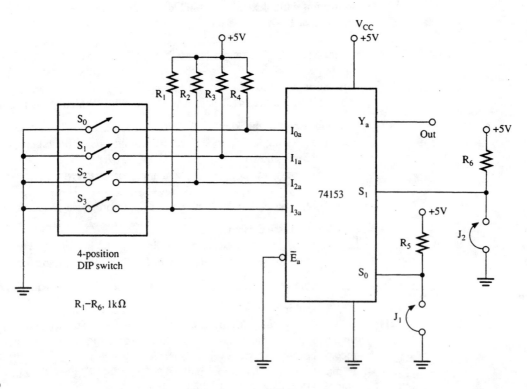

FIGURE 9.2

Part 3: Circuit Applications

Polling Input Lines: Watchdog Multiplexer Circuit

☐ 1. Multiplexers are often used to ''poll'' data-in lines. In such cases, each input is looked at in sequence for a brief instant. Data present at the time of selection is transferred to the output. Depending on the size of the multiplexer and how many are cascaded, hundreds, even thousands of input lines can be scanned, with a tremendous saving in circuit interconnections and wiring.

A circuit designed to illustrate this concept is shown in Figure 9.3. A 1-kHz clock drives a 7493 binary counter, which in turn selects, or addresses, select inputs S_0 and S_1 of a 74153. The select inputs cycle rapidly through a count of 00, 01, 10, and 11. As they do, data-in lines I_{0a}–I_{3a} are polled. Thus each input is looked at 250 times a second (1000 Hz divided by four inputs).

As long as all inputs are LOW (switches S_0–S_3 are closed), output Y_a remains LOW. When any switch, S_0–S_3, is opened, bringing its respective input HIGH, output Y_a will go HIGH at the time the switch is polled. In this Watchdog Multiplexer Circuit, a HIGH at Y_a, even for only a split second, triggers the SCR into conduction and latches the LED (load) on. Pressing S_4 resets the SCR.

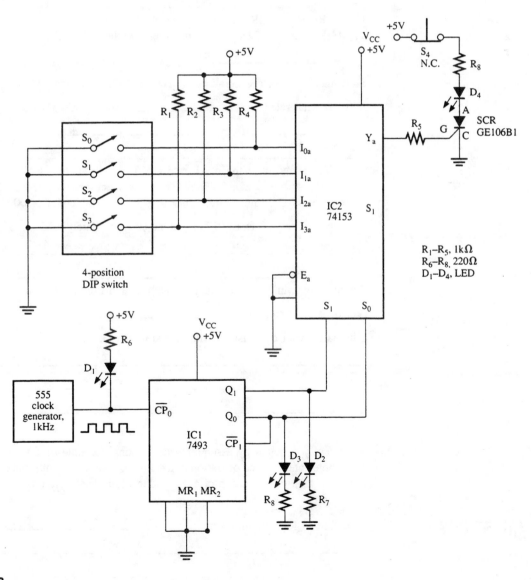

FIGURE 9.3

While this circuit may seem overly complex for monitoring just four input switches, keep in mind that its purpose is to illustrate the polling concept. Its design advantage—multiplexing—really comes through when a large number of inputs are inspected.

☐ 2. Referring to Figure 9.3, record missing IC pin numbers. Construct the circuit of Figure 9.3. Apply power. Place switches S_0–S_3 LOW. Do LEDs D_1, D_2, and D_3

   appear to be on all the time? _____ Are they actually on continuously,

   or are they flashing at a rate too fast for the human eye to detect? Why is this so?

   _____

   _____

   Is LED D_4 on or off? _____ Why is this so? _____

   _____

   ☐ a. Now open any switch, S_0–S_3. What happens to LED D_4? _____

      Why is this so? _____

      Does LED D_4 appear to come on immediately after an input switch is

      opened? _____ Why is this so? _____

      _____

      Reset the SCR circuit by pressing S_4.

   ☐ b. Repeat step 2a for the remaining switches, S_0–S_3. Are the results the

      same? _____

☐ 3. Slow the 555 clock generator to 1 Hz. Is LED D_1 flashing? _____

   If so, at what rate? _____ Are LEDs D_2 and D_3 flashing?

   _____ If so, why? _____

   _____

   We have slowed the clock rate to 1 Hz in order to "see" polling as it actually takes place. LEDs D_2 and D_3 monitor select inputs S_0 and S_1. Keep your eye on D_2 and D_3. Immediately *after* they indicate the count of 00 (both LEDs

   off), open switch S_0. Does LED D_4 come on immediately? _____ If so,

   why? _____

   _____

   Try this step while opening another input switch immediately after it has been

   polled. Is the result the same? _____

☐ 4. Summarize your conclusions with regard to step 3: _____

   _____

   _____

   _____

SUMMARY

In this experiment you began by assembling and testing a simple discrete multiplexer. Next you assembled and verified operation of a 4-line-to-1-line MSI multiplexer. Finally, you built and analyzed the operation of a Watchdog Multiplexer Circuit.

REVIEW QUESTIONS

1. Write a definition for a multiplexer. _____

   _____

   _____

2. With regard to the 2-line-to-1-line multiplexer of Figure 9.1, explain how input-line selection takes place. _____

_____

_____

3. Explain the concept of polling as it applies to a multiplexer circuit. _____

_____

4. Why do you suppose an SCR, not a transistor, is used in the circuit of Figure 9.3?

_____

_____

_____

Writing Skills Assignment

In as few paragraphs as possible, explain how the Watchdog Multiplexer Circuit works. You might first discuss what the circuit does and then discuss how it does it.

10 Demultiplexers

OBJECTIVES

After completing this experiment, you will be able to

- Assemble and verify operation of a discrete demultiplexer
- Verify operation of a dual 2-line-to-4-line demultiplexer
- Build and analyze a Four-Position Monitor Circuit

REFERENCE READING

Review Ronald Reis, *Digital Electronics Through Project Analysis*, Chapter 5, Section 5.1, and Chapter 8, Section 8.1.

EQUIPMENT & MATERIALS NEEDED

Equipment

- [ ] 1   5-V power supply
- [ ] 1   logic probe
- [ ] 1   logic pulser
- [ ] 1   555 clock generator
- [ ] 1   solderless circuit board

Materials

- [ ] 1   7404 hex inverter
- [ ] 1   7408 quad two-input AND gate
- [ ] 1   7493 binary counter
- [ ] 1   74153 dual 4-line-to-1-line multiplexer
- [ ] 1   74155 2-line-to-4-line demultiplexer
- [ ] 7   LEDs (red)
- [ ] 4   220-Ω resistors
- [ ] 4   1-kΩ resistors
- [ ] 1   8-position DIP switch (or equivalent)
- [ ] 1   package of jumper wires

BACKGROUND INFORMATION

A demultiplexer is the reverse of a multiplexer. It has one data-input line and several data-output lines. Data arriving on the input is channeled to a selected output. Hence a demultiplexer is also referred to as a *data distributor*.

There are 2-, 4-, 8-, and 16-line demultiplexers available in MSI DIP packages, both TTL and CMOS. In this experiment you will begin by building a discrete 1-line-to-2-line DEMUX to examine how data-line distribution takes place. Next you will test the TTL 74155, a dual 2-line-to-4-line demultiplexer. Finally, you will use both the 74153 multiplexer and the 74155 demultiplexer to build a Four-Position Monitor Circuit that illustrates the complete data selection and distribution process.

PROCEDURE

Part 1: Circuit Fundamentals

A Discrete Demultiplexer

☐ 1. A simple discrete 1-line-to-2-line demultiplexer using two AND and three NOT gates is shown in Figure 10.1. As the accompanying truth table indicates, when select output (address) line S is LOW, data entering on the data-in line will be channeled to output line A. If the data entering is LOW, the data emerging on line A will be LOW. If data entering is HIGH, line A will go HIGH. Output B will remain LOW because one of the inputs to AND gate B (the lower input line) is always LOW.

When select output line S is placed HIGH, the reverse takes place; now data arriving at the input is channeled to output line B. As in the previous case, data emerging is in true (noninverted) form. Line A now remains LOW because one of the inputs to AND gate A (the lower input line) is always LOW.

   ☐ a. Referring to Figure 10.1, record missing IC pin numbers. Construct the circuit in Figure 10.1 Apply power.

   ☐ b. Place select output line S LOW, as indicated in step 1 of Table 10.1. Using a logic pulser, or the 555 clock generator set to 5 Hz, place a signal on the data-in line. Using a logic probe, measure the logic level on output line A. Record the results—a LOW, a HIGH, or a pulse—in the space

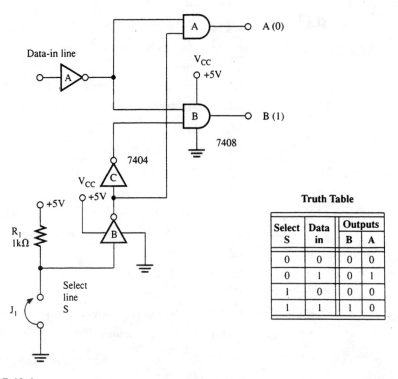

Truth Table

Select S	Data in	Outputs	
		B	A
0	0	0	0
0	1	0	1
1	0	0	0
1	1	1	0

FIGURE 10.1

TABLE 10.1

Step	Select s	Data in	Outputs B	A
1	0	⊓⊔⊓⊔		
2	1	⊓⊔⊓⊔		

provided. Next measure the logic level on output line B. Record the result.

☐ c. Perform step 2 of Table 10.1 in a similar manner, recording your results in the space provided. Were your results as anticipated? _____

Part 2: Further Investigation

An MSI Demultiplexer

☐ 1. To gain experience with a demultiplexer chip, and at the same time provide one that is compatible with the 74153 MUX, we have selected the TTL 74155 2-line-to-4-line demultiplexer, the pin configuration for which is shown in Appendix A (Figure A.24). There are two 2-line-to-4-line demultiplexers in each IC, designated a and b. Each has its own active-LOW enable pin—$\overline{E}_a$ (pin 2) for DEMUX a and $\overline{E}_b$ (pin 14) for DEMUX b—as well as its own group of four data-out lines: $\overline{0}_a$, $\overline{1}_a$, $\overline{2}_a$, and $\overline{3}_a$ for DEMUX a; $\overline{0}_b$, $\overline{1}_b$, $\overline{2}_b$, and $\overline{3}_b$ for DEMUX b. The input for DEMUX a is E_a (pin 1); the input for DEMUX b is $\overline{E}_b$ (pin 15). Both demultiplexers, however, share the same select, or address, lines, A_0 and A_1. It should be noted that DEMUX a inverts at its outputs data applied to the input. DEMUX b does not; data arriving at its input appears in true form (noninverted) at the outputs. Finally, ground is at pin 8; V_{CC}, at pin 16.

☐ 2. A test circuit for the 74155 is shown in Figure 10.2. As you can see, DEMUX $\overline{0}_b$ is the one being used.

☐ a. Referring to Figure 10.2, record missing IC pin numbers. Construct the circuit shown in Figure 10.2. Apply power.

☐ b. To check chip operation, set up the input conditions shown in Table 10.2.

Select		Data in (pin 15)	Enable (pin 14)	Outputs			
A_0	A_1	$\overline{E}_b$	$\overline{E}_b$	$\overline{0}_b$	$\overline{1}_b$	$\overline{2}_b$	$\overline{3}_b$
0	0	0	0	0	1	1	1
0	0	1	0	1	1	1	1
1	0	0	0	1	0	1	1
1	0	1	0	1	1	1	1
0	1	0	0	1	1	0	1
0	1	1	0	1	1	1	1
1	1	0	0	1	1	1	0
1	1	1	0	1	1	1	1

TABLE 10.2

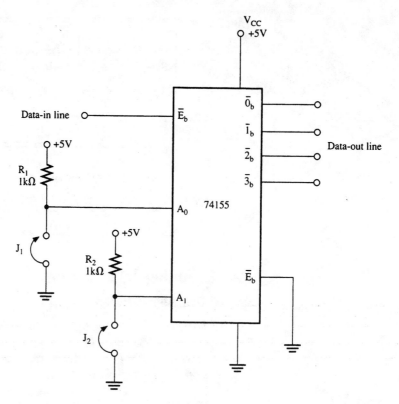

FIGURE 10.2

As you select each output line with jumpers J_1 and J_2, place a LOW or HIGH on data-in line $\overline{E}_b$ (pin 15), as shown in the table. Using a logic probe, measure the output logic level on all four outputs. Record the results in the space provided. Summarize your results: _____

_____

_____

_____

☐ 3. Repeat step 2b, only this time use a logic pulser, or the 555 clock generator set to 5 Hz, to place a signal on data-in line $\overline{E}_b$ (pin 15). Record the output results—a LOW, a HIGH, or a pulse—for each setting of select lines A_0 and A_1 in the space provided in Table 10.3. Summarize your results: _____

_____

_____

_____

Part 3: Circuit Applications

Data Selection and Distribution: Four-Position Monitor Circuit

☐ 1. The circuit of Figure 10.3 illustrates how a multiplexer and demultiplexer can team up to select, transmit, and distribute data. The bulk of the circuit— consisting of the 555 clock, the 7493, the 74153, and associated components— we have seen before, in Experiment 9, Figure 9.3. With the clock frequency set to 1 kHz, each data-in line to IC2 is scanned, or polled, 250 times a second. If any switch, S_0–S_3, opens, output Y_a of IC2 goes HIGH.

The circuit of Figure 10.3, unlike that of Figure 9.3, also contains IC4, the 74155 demultiplexer. Since IC4's select lines A_0 and A_1 are paralleled with select

Select		Data in (pin 15)	Enable (pin 14)	Outputs			
A_0	A_1	$\overline{E}_b$	$\overline{E}_b$	$\overline{0}_b$	$\overline{1}_b$	$\overline{2}_b$	$\overline{3}_b$
0	0	⎍⎍	0				
1	0	⎍⎍	0				
0	1	⎍⎍	0				
1	1	⎍⎍	0				

TABLE 10.3

lines S_0 and S_1 of IC2, data-out lines $\overline{0}_b$–$\overline{3}_b$ are scanned in synchronization with the polling of data-in lines I_{0a}–I_{3a}. Thus when an input switch, S_0–S_3, opens, the corresponding LED (D_4–D_7) lights. As you can see, the HIGH signal on the Y_a output of IC2 is inverted by NOT gate A of IC3, then sent to the active-LOW input of IC4 ($\overline{E}_b$). The incoming LOW at $\overline{E}_b$ is channeled to the corresponding output, and the respective LED turns on.

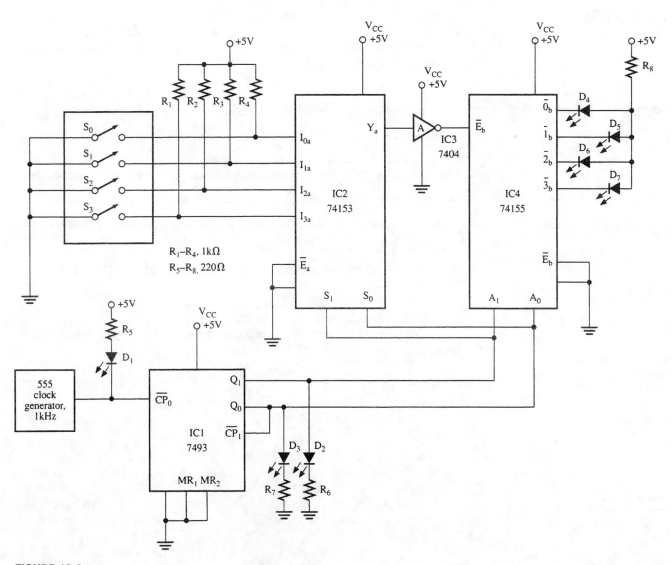

FIGURE 10.3

87

☐ 2. Referring to Figure 10.3, record missing IC pin numbers. Construct the circuit of Figure 10.3. Apply power. Close all input switches, S_0–S_3. Do any output LEDs, D_4–D_7, light? _____ Why is this so? _____

    ☐ a. Open switch S_0. Does any output LED light? _____ If so, which one? _____ Why is this so? _____

    ☐ b. Open two input switches, S_0 and S_1. Which output LEDs light? _____ Why is this so? _____

       Are the lit LEDs on continuously, or do they just appear to be? _____ Why is this so? _____

    ☐ c. Proceed to check out the remaining inputs. Do corresponding LEDs light as they should? _____

SUMMARY

In this experiment you began by assembling and testing a simple discrete demultiplexer. Next you assembled and verified operation of a 2-line-to-4-line MSI demultiplexer. Finally, you built and analyzed operation of a Four-Position Monitor Circuit incorporating both a multiplexer and demultiplexer.

REVIEW QUESTIONS

1. Write a definition for a multiplexer. _____

2. With regard to the 1-line-to-2-line demultiplexer of Figure 10.1, explain how output-line selection takes place. _____

3. What is the difference between the two demultiplexers inside the 74155 IC? _____

4. Redesign the circuit of Figure 10.3 to eliminate the need for NOT gate A. (*Hint:* Look over your answer to Question 3.)

Writing Skills Assignment

In as few paragraphs as possible, explain how the Four-Position Monitor Circuit works. You might first discuss what the circuit does and then how it does it.

11 Adders

OBJECTIVES

After completing this experiment, you will be able to

- Construct and then verify operation of a discrete half adder
- Construct and then verify operation of a discrete full adder
- Build and analyze a Full-Adder Demonstration Circuit using the TTL MSI 7483 IC

REFERENCE READING

Review Ronald Reis, *Digital Electronics Through Project Analysis*, Chapter 5, Section 5.2.

EQUIPMENT & MATERIALS NEEDED

Equipment

- ☐ 1 5-V power supply
- ☐ 1 solderless circuit board

Materials

- ☐ 1 7408 quad two-input AND gate
- ☐ 1 7432 quad two-input OR gate
- ☐ 1 7483 4-bit full adder
- ☐ 1 7486 quad two-input exclusive-OR gate
- ☐ 5 LEDs (red)
- ☐ 5 220-Ω resistors
- ☐ 8 1-kΩ resistors
- ☐ 1 8-position DIP switch (or equivalent)
- ☐ 1 package of jumper wires

BACKGROUND INFORMATION

To perform complete binary addition, you need a logic circuit that will accept three input digits (an augend, addend, and a carry in from a previous column) and supply two output digits (a sum and a carry out to the next column). Such a circuit is known as a *full adder*, the block diagram of which is shown in Figure 11.1a. A full adder, however, is made up of two *half adders*. A half adder will supply two output digits (sum and carry) but accept

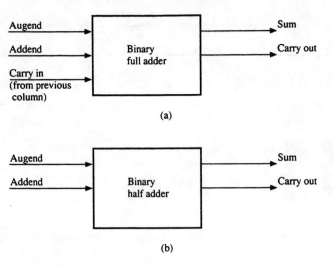

FIGURE 11.1   (a) Full-adder circuit and (b) half-adder circuit.

only two input digits (augend and addend). It cannot receive a carry in from a previous column (Figured 11.1b).

In this experiment you will begin by constructing and then analyzing the operation of a discrete half adder. Next you will do the same with a discrete full adder made up of two half adders plus a two-input OR gate. Third, you will examine the TTL 7483 4-bit full adder MSI chip that will add two 4-bit words. You will build and test a Full-Adder Demonstration Circuit using the 7483.

PROCEDURE

Part 1: Circuit Fundamentals

Half-Adder Operation

☐   1. A simple discrete half adder can be constructed using one two-input AND gate and one two-input XOR gate, as shown in Figure 11.2. LED D_1 indicates a sum; LED D_2, a carry out. Inputs A and B represent the augend and addend, respectively. Trace through any binary input combination (00, 01, 10, 11) to see if the circuit indeed adds. *Note:* A lit LED indicates a HIGH (1) output, a dark LED, a LOW (0) output.

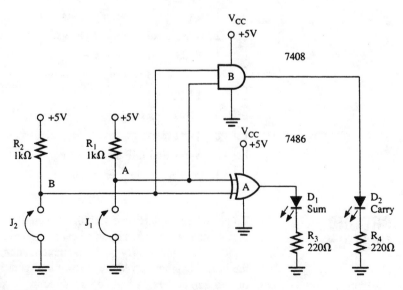

FIGURE 11.2

TABLE 11.1

B	A	Carry	Sum
0	0		
0	1		
1	0		
1	1		

☐ a. Referring to Figure 11.2, record missing IC pin numbers. Construct the circuit shown in Figure 11.2. Apply power.

☐ b. Set up the input combinations shown in Table 11.1 and record the output logic levels in the space provided. Are your results consistent with those anticipated? _____

Part 2: Further Investigation

Full-Adder Operation

☐ 1. With two half adders and a two-input OR gate, you can construct a full adder—a circuit that will accept augend, addend, and carry-in digits and produce sum and carry-out digits. The block diagram for a full adder is shown in Figure 11.1a; the complete schematic, in Figure 11.3. With enough full adders cascaded, two binary words of any length can be added.

    ☐ a. Referring to Figure 11.3, record missing IC pin numbers. Construct the circuit in Figure 11.3. Apply power.

    ☐ b. The input combinations shown in Table 11.2 represent all input possibilities (augend, addend, and carry-in digits) for a full adder. Set up the combinations shown and record the output logic levels in the space provided. How many input combinations result in both LEDs being lit?

_____

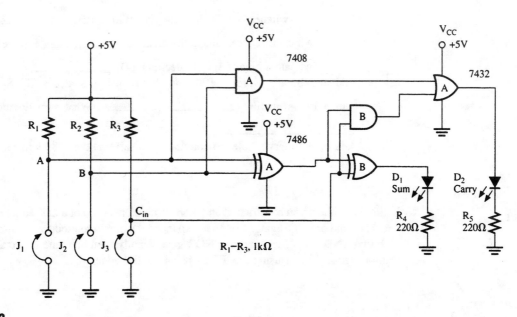

FIGURE 11.3

TABLE 11.2

B	A	C_{in}	Carry	Sum
0	0	0		
0	0	1		
0	1	0		
0	1	1		
1	0	0		
1	0	1		
1	1	0		
1	1	1		

☐ 3. Summarize your results with regard to Part 2: _____

_____

_____

_____

Part 3: Circuit Applications

An MSI Full-Adder IC: Full-Adder Demonstration Circuit

☐ 1. The 7483 is a TTL 16-pin chip that will add two 4-bit binary words (A_n plus B_n) plus the incoming carry. The binary sum appears on the output, S_1–S_4, and the outgoing carry, C_4. The pin configuration for the 7483 is shown in Appendix A (Figure A.11). A test circuit is shown in Figure 11.4.

   ☐ a. Referring to Figure 11.4, record missing IC pin numbers. Construct the circuit shown in Figure 11.4. Apply power.

   ☐ b. To test the circuit, add binary words 1010 and 1001 (decimal 10 and decimal 9). Input word A will be the augend; input word B, the addend.

   Set switches S_1–S_8 accordingly. Which LEDs are lit? _____

   Which LEDs remain dark? _____ Does the output

   indicate a binary 10011 (decimal 14)? _____

☐ c. Add the binary words 1100 and 1111, using switches S_1–S_8 as inputs.

   Which LEDs are lit? _____ What sum do they indicate?

   _____

☐ d. Add other binary nibbles to convince yourself that the 7483 will indeed add any two 4-bit binary words.

SUMMARY

In this experiment you constructed and then verified operation of a half-adder circuit using one AND and one XOR gate. You then expanded the circuit to produce a full-adder circuit with two AND, two XOR, and one NOR gate. Finally, you built and tested a Full-Adder Demonstration Circuit using the TTL 7483 4-bit full-adder IC.

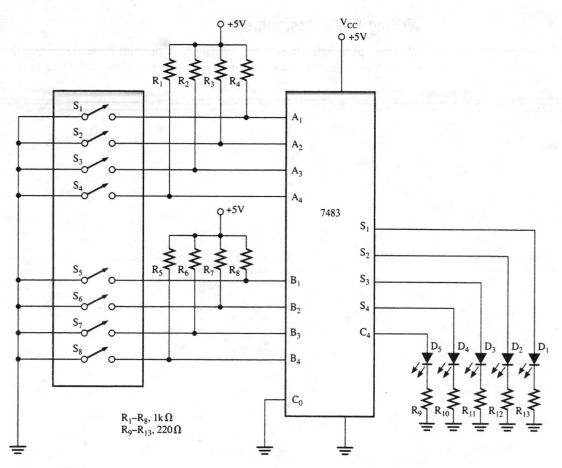

FIGURE 11.4

REVIEW QUESTIONS

1. Explain the difference between a half adder and full adder. _____
_____
_____

2. How many full adders are required to add two numbers with six digits each?
_____

3. Does the 7483 have active-LOW or active-HIGH outputs?
4. Draw the schematic for an 8-bit binary adder using two 7483 ICs.

Writing Skills Assignment

In as few paragraphs as possible, explain how the Full-Adder Demonstration Circuit works. You might first discuss what the circuit does and then how it does it.

12 Astable Multivibrators

OBJECTIVES

After completing this experiment, you will be able to

- Assemble an astable multivibrator using a 555 IC
- Vary the frequency and duty cycle of an astable multivibrator
- Use an oscilloscope to measure the frequency and duty cycle of an astable multivibrator
- Build and analyze a Windshield-Wiper-Fluid-Level Detector Circuit
- Build and analyze an Audio-Continuity Checker Circuit

REFERENCE READING

Review Ronald Reis, *Digital Electronics Through Project Analysis*, Chapter 9, Section 9.1.

EQUIPMENT & MATERIALS NEEDED

Equipment

- [ ] 1 5-V power supply
- [ ] 1 dual-channel oscilloscope
- [ ] 1 solderless circuit board

Materials

- [ ] 1 555 timer
- [ ] 1 LED (red)
- [ ] 1 0.01-μF capacitor
- [ ] 1 0.1-μF capacitor
- [ ] 1 10-μF capacitor
- [ ] 1 47-μF capacitor
- [ ] 1 100-μF capacitor
- [ ] 1 220-Ω resistor
- [ ] 1 1-kΩ resistor
- [ ] 1 4.7-kΩ resistor
- [ ] 1 10-kΩ resistor
- [ ] 1 22-kΩ resistor
- [ ] 1 100-kΩ resistor

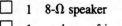
□ 1 8-Ω speaker
□ 1 package of jumper wires

BACKGROUND INFORMATION

The astable multivibrator is an oscillator, or signal generator, with no stable state. Its output voltage alternates between LOW and HIGH at a constant frequency. That frequency is usually determined by an R/C time constant. Such a circuit is also known as a *clock,* or *free-running oscillator*.

The pin configuration for a 555 timer IC is shown in Appendix A (Figure A.28). The IC, configured as an astable multivibrator, is shown in Figure 12.1. The output frequency and duty cycle are determined by the values of R_1, R_2, and C_1.

The output is HIGH during t_1, when C_1 is charging to V_{CC} through R_1 and R_2.

The output is LOW during t_2, when C_1 is discharging to ground through R_2.

As we can see, $t_1 + t_2$ equals the period T of the output signal. Since frequency f is the reciprocal of the period, $f = 1/T$.

In this experiment you will first learn how to determine the output frequency of a 555 astable multivibrator. Then you will discover how to determine the output duty cycle for such a circuit. And finally, you will build and analyze two practical astable multivibrator circuits: the Windshield-Wiper-Fluid-Level Detector Circuit and the Audio-Continuity Checker Circuit.

PROCEDURE

Part 1: Circuit Fundamentals

Determining Output Frequency of a 555 IC Astable Multivibrator

□ 1. With a 555 IC, T is determined by the formula:

$$T = 0.693(R_1 + 2R_2)C_1$$

And frequency f is

$$f = \frac{1}{0.693(R_1 + 2R_2)C_1}$$

which is the reciprocal of T.

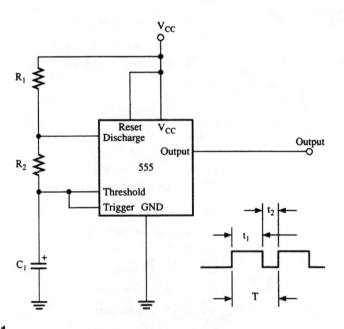

FIGURE 12.1

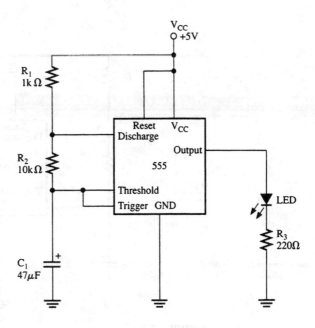

FIGURE 12.2

For example, if R_1 is $1\,k\Omega$, R_2 is $10\,k\Omega$, and C_1 is $47\,\mu F$, the frequency is

$$f = \frac{1}{0.693(1000 + 20,000)0.000047}$$
$$= 1.46\ Hz$$

☐ 2. Referring to Figure 12.2, record missing IC pin numbers. Construct the circuit shown in Figure 12.2. Apply power. The LED should flash at approximately 1.46 Hz. Does it? _____

Since output frequency is inversely proportional to R_1, R_2, and C_1, increasing C_1 will reduce the frequency; decreasing C_1 will increase the frequency.

☐ a. Replace the 47-μF capacitor with a 100-μF capacitor. Does the LED flash rate (frequency) increase or decrease? _____ Why is this so? _____

_____

☐ b. Replace the 100-μF capacitor with a 10-μF capacitor. Does the LED flash rate increase or decrease compared with the original capacitance value? _____ Why is this so? _____

_____

_____

Part 2: Further Investigation

Determining Output Duty Cycle of a 555 IC Astable Multivibrator

☐ 1. With a 555 IC, duty cycle D is determined by the formula

$$D = \frac{t_1}{T} = \frac{R_1 + R_2}{R_1 + 2R_2} \times 100$$

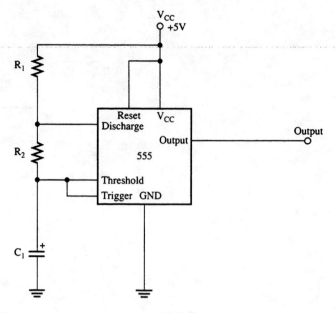

FIGURE 12.3

For example, if R_1 is 2.2 kΩ and R_2 is 680 kΩ, the duty cycle is

$$D = \frac{2200 + 6800}{2200 + 13{,}600} \times 100$$
$$= 57\%$$

2. Referring to Figure 12.3, record missing IC pin numbers. Construct the circuit shown in Figure 12.3. Use a 4.7-kΩ resistor for R_1, a 10-kΩ resistor for R_2, and a 0.01-μF capacitor for C_1. Apply power.

   a. Calculate the frequency and duty cycle:

      f = _____; D = _____

   b. Using an oscilloscope, observe the waveform of the 555 output. Draw the waveform on Plot 12.1. Determine the observed frequency and duty cycle:

      f = _____; D = _____

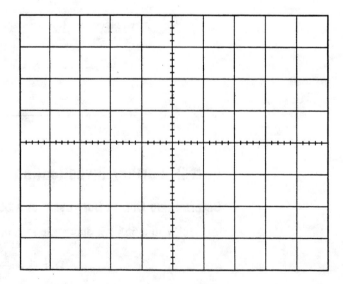

PLOT 12.1

☐ 3. Repeat step 2, using an R_1 of $10\,k\Omega$, an R_2 of $22\,k\Omega$, and a C_1 of $0.01\,\mu F$.

   ☐ a. Calculate the frequency and duty cycle:

      $f =$ _____; $D =$ _____

   ☐ b. Using an oscilloscope, observe the waveform of the 555 output. Draw the waveform on Plot 12.2. Determine the observed frequency and duty cycle:

      $f =$ _____; $D =$ _____

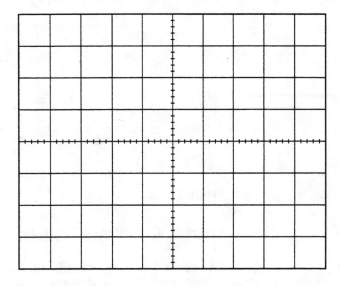

PLOT 12.2

☐ 4. Repeat step 2, using an R_1 of $1\,k\Omega$, an R_2 of $22\,k\Omega$, and a C_1 of $0.1\,\mu F$.

   ☐ a. Calculate the frequency and duty cycle:

      $f =$_____; $D =$ _____

   ☐ b. Using an oscilloscope, observe the waveform of the 555 output. Draw the waveform on Plot 12.3. Determine the observed frequency and duty cycle:

      $f =$ _____; $D =$ _____

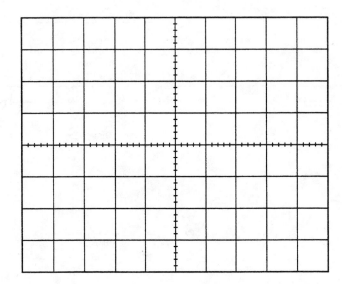

PLOT 12.3

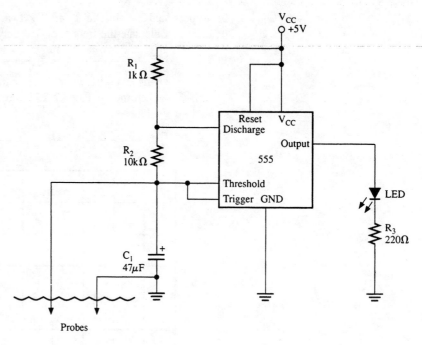

FIGURE 12.4

Part 3: Circuit Applications

A Practical Astable Multivibrator Circuit: Windshield-Wiper-Fluid-Level Detector Circuit

☐ 1. There are hundreds of applications for a clock, or astable multivibrator. A practical circuit for your car is shown in Figure 12.4. When the two probes (wires) of the Windshield-Wiper-Fluid-Level Detector Circuit are immersed in water, C_1 is shorted, no capacitor charging takes place, and the oscillator does not function. When the fluid level is below the probes, C_1 is free to charge and discharge, and the LED "warning light" flashes at a 1.46-Hz rate.

    ☐ a. Referring to Figure 12.4, record missing IC pin numbers. Construct the circuit shown in Figure 12.4. (*Note:* While in this experiment the circuit

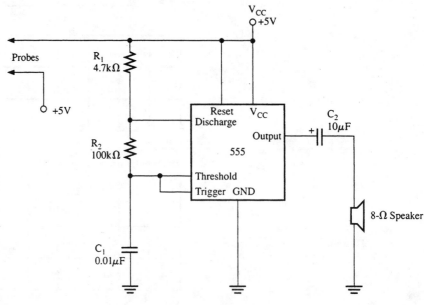

FIGURE 12.5

102

is operated from a 5-V power supply, the device will work just as well with 12-V from your car.) Apply power.

☐   b. Test the circuit by immersing the two probes in a cup of water. The LED should not flash. Remove the probes and the LED will flash. Does it?

A Practical Astable Multivibrator Circuit: Audio-Continuity Checker Circuit

☐   1. A simple audio-continuity tester can be built using a 555 IC astable multivibrator. The circuit is shown in Figure 12.5. When a conducting material is placed between the probes, a tone will be heard from the speaker. With the probes open, there is no continuity, no power to the IC, and no oscillation.

☐   a. Referring to Figure 12.5, record missing IC pin numbers. Construct the circuit shown in Figure 12.5. Apply power.

☐   b. Test the circuit by touching the probes together. Do you hear a tone from the speaker? _____

SUMMARY

In this experiment you assembled an astable multivibrator, varied frequency and duty cycle, and constructed two applications circuits: a Windshield-Wiper-Fluid-Level Detector and an Audio-Continuity Checker.

REVIEW QUESTIONS

1. The frequency of an astable multivibrator is usually determined by an _____ time constant.

2. A 555 IC astable multivibrator has an R_1 of $10\,k\Omega$, an R_2 of $100\,k\Omega$, and a C_1 of $0.22\,\mu F$. What is the circuit's output frequency? _____ What is its duty cycle? _____

3. A 555 IC astable multivibrator has an R_1 of $100\,k\Omega$, an R_2 of $470\,k\Omega$, and a C_1 of $1\,\mu F$. What is the circuit's output frequency? _____ What is its duty cycle? _____

4. A 555 IC astable multivibrator has an R_1 of $68\,k\Omega$, an R_2 of $120\,k\Omega$, and a C_1 of $1\,\mu F$. What is the circuit's output frequency? _____ What is its duty cycle? _____

Writing Skills Assignment

In as few paragraphs as possible, explain how either the Windshield-Wiper-Fluid Detector Circuit or the Audio-Continuity Checker Circuit works. You might first discuss what the circuit does and then discuss how it does it.

13 Monostable Multivibrators

OBJECTIVES

After completing this experiment, you will be able to

- Assemble a monostable multivibrator using a 555 IC
- Vary the pulse width of a 555 monostable multivibrator
- Examine reset and retrigger options with a 555 monostable multivibrator
- Examine positive and negative edge triggering, complementary outputs, and short-duration pulse creation with the 74121 chip
- Build and analyze a Dark/Light-Activated Timer Circuit

REFERENCE READING

Review Ronald Reis, *Digital Electronics Through Project Analysis*, Chapter 9, Section 9.2.

EQUIPMENT & MATERIALS NEEDED

Equipment

- ☐ 1 5-V power supply
- ☐ 1 dual-channel oscilloscope
- ☐ 1 solderless circuit board

Materials

- ☐ 1 555 timer
- ☐ 1 74121 monostable multivibrator
- ☐ 1 LED (red)
- ☐ 1 LED (green)
- ☐ 1 0.01-μF capacitor
- ☐ 1 0.1-μF capacitor
- ☐ 1 4.7-μF capacitor
- ☐ 1 10-μF capacitor
- ☐ 1 47-μF capacitor
- ☐ 1 470-μF capacitor
- ☐ 2 220-Ω resistors
- ☐ 3 1-kΩ resistors

- ☐ 2 10-kΩ resistors
- ☐ 1 18-kΩ resistor
- ☐ 1 1-MΩ resistor
- ☐ 1 2.2-MΩ resistor
- ☐ 2 photoresistors
- ☐ 3 N.O. push-button switches
- ☐ 1 package of jumper wires

BACKGROUND INFORMATiON

The monostable multivibrator is a circuit whose output has one stable state, usually LOW. When the circuit is externally triggered, its output goes to the unstable state (HIGH) for a predetermined time, after which it returns to its stable state (Figure 13.1). An R/C time constant determines the time for the unstable state. Monostable multivibrators are often called *timers* since they provide an output pulse for a given duration (time). They are also referred to as "one shots," because when triggered, they output one pulse, or one "shot."

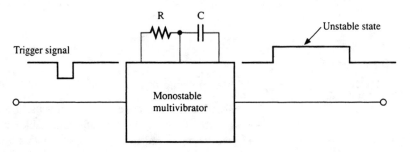

FIGURE 13.1

The 555 is the premier timer IC. It can produce a HIGH output pulse from as short a time as 1 μs to as long as 15 min simply by adjusting the value of a single resistor and a single capacitor. In this experiment you will create a monostable multivibrator using the 555 IC. You will then select various resistor and capacitor values so as to alter the output pulse time. In addition, you will examine the effects of reset and retriggering on this ubiquitous one-shot circuit.

As popular as the 555 IC is, it cannot meet all timer needs. If you demand extremely short output pulse widths (down to several tens of nanoseconds), multiple inputs with both positive and negative edge triggering, or complementary outputs, the TTL 74121 is a logical choice. In this experiment you will examine how this 14-pin chip accomplishes all of the above.

Finally, using the 555 again, you will build and analyze a Dark/Light-Activated Timer Circuit, which has countless practical applications.

PROCEDURE

Part 1: Circuit Fundamentals

Determining Monostable Output Times for a 555 IC

☐ 1. The 555 IC, the pin configuration of which is shown in Appendix A (Figure A.28) is the most popular timer IC ever made. Knowing more about its operation is fundamental to your work in digital electronics.

Determining monostable output times for the 555 IC is easy; the formula is

$$t = 1.1 \times R_1 \times C_1$$
where t = time, in seconds
1.1 = a constant
R_1 = resistance, in ohms
C_1 = capacitance, in farads

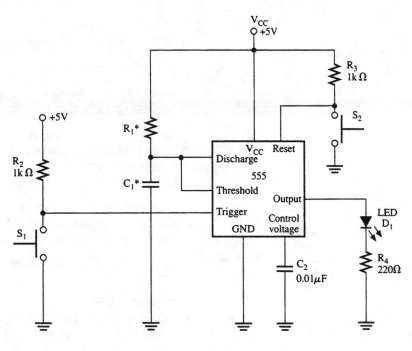

* See table 13.1.

FIGURE 13.2

For example, if R_1 is 470 kΩ and C_1 is 47 μF, the output pulse width, or duration, is

$$t = 1.1 \times R_1 \times C_1$$
$$= 1.1 \times 470,000 \times 0.000047$$
$$= 24.29 \text{ s}$$

☐ 2. Examine the monostable multivibrator circuit of Figure 13.2. Note that the trigger input is pulled HIGH through pull-up resistor R_2 and that the reset pin is pulled HIGH through pull-up resistor R_3. Resistor R_1 and capacitor C_1 form the R/C time constant. The red LED will light when the output goes HIGH.

    ☐ a. Referring to Figure 13.2, record missing IC pin numbers. Construct the circuit of Figure 13.2 using the values of R_1 and C_1 in row 1 of Table 13.1.

    ☐ b. Using the formula presented above, calculate the output pulse duration and record your answer in the space provided in row 1 of Table 13.1.

    ☐ c. Apply power. Momentarily press switch S_1, bringing the trigger input LOW and thus triggering the monostable multivibrator. Does the LED come on? _____ If so, after the LED goes out, press S_1 again, this time measuring the LED on time. Record the measured result in

	R_1	C_1	Output	Pulse
			Calculated	**Measured**
Row 1	1 MΩ	4.7 μF		
Row 2	1 MΩ	10 μF		
Row 3	2.2 MΩ	10 μF		
Row 4	1 MΩ	47 μF		

TABLE 13.1

the space provided in Table 13.1. With allowances for resistor and capacitor tolerance variations, are the calculated and measured values consistent? _____

☐ d. Repeat all the parts of step 2, using the resistor and capacitor values in rows 2, 3, and 4 of Table 13.1. In each case, calculate and record the output pulse duration first. Then measure and record the pulse width. Are the calculated and measured results consistent? _____

☐ 3. The 555 monostable multivibrator can be reset at any time after the ouput goes HIGH, by momentarily bringing pin 4 LOW.

☐ a. In the circuit of Figure 13.2, make R_1 1 MΩ and C_1 10 μF. Press S_1. The pulse duration should be close to 11 s. Is it? _____

☐ b. Now press S_1 again, starting the timing cycle. Next, press S_2 any time after the timing cycle begins. Does the LED go off when S_2 is pressed? _____ Repeat the above procedure a few times, pressing S_1 first, then some time later, S_2. Does the circuit always reset when the reset pin is brought LOW? _____

☐ 4. With the 555 timer, if additional trigger pulses occur when the output is HIGH, there is no effect on the output pulse time. In other words, the pulse width is not lengthened.

☐ a. Using the previous circuit, press S_1, starting the timing cycle of 11 s. Press S_1 again a few seconds later. Does the timing cycle increase in length? _____

☐ b. Repeat step 4a a few times to convince yourself that the circuit cannot be retriggered once an initial trigger takes place.

Part 2: Further Investigation

A TTL Monostable Multivibrator

☐ 1. The TTL 74121 IC, the pin configuration of which is shown in Appendix A (Figure A.17), is a monostable multivibrator featuring both positive and negative edge triggering and complementary outputs. Pulse width, determined by an R/C time constant, can vary from 30 ns to 28 s. There are three trigger inputs: two negative edge ($\overline{A}_1$ and $\overline{A}_2$) and one positive edge (B).

Determining monostable output times for the 74121 timer is easy; the formula is

$$t = 0.693 \times R_1 \times C_1$$

where t = time, in seconds
0.693 = a constant
R_1 = resistance, in ohms
C_1 = capacitance, in farads

(*Note:* Resistor R_1 can range from 2 to 40 kΩ and the capacitor from 10 pF upward.)

For example, if R_1 is 22 kΩ and C_1 is 4.7 μF, the output pulse width, or duration, is

$$\begin{aligned} t &= 0.693 \times R_1 \times C_1 \\ &= 0.693 \times 22{,}000 \times 0.0000047 \\ &= 716 \text{ ms} \end{aligned}$$

☐ 2. Examine the monostable multivibrator circuit of Figure 13.3. Note that trigger inputs $\overline{A}_1$, $\overline{A}_2$, and B are pulled HIGH through respective pull-up resistors R_2, R_3, and R_4. Pressing switch S_1, thus causing a HIGH-to-LOW transition (negative edge trigger) on input $\overline{A}_1$, while leaving inputs $\overline{A}_2$ and B HIGH,

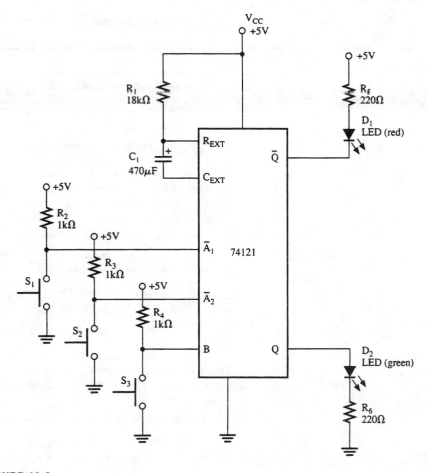

FIGURE 13.3

triggers the one-shot. As a result, output $\overline{Q}$ goes LOW and output Q goes HIGH for the timing duration. See Table 13.2. Pressing switch S_2, causing a similar HIGH-to-LOW transition on input $\overline{A}_2$, while leaving inputs $\overline{A}_1$ and B HIGH, also triggers the timer, with similar output results.

Finally, to create an output pulse with a LOW-to-HIGH transition (positive edge trigger), input B is used. In this case, either inputs $\overline{A}_1$ or $\overline{A}_2$ (or both) must be held LOW (Table 13.2).

☐ a. Referring to Figure 13.3, record missing IC pin numbers. Construct the circuit of Figure 13.3. Apply power.

☐ b. Calculate the output pulse duration using an R_1 of 18 kΩ and a C_1 of 470 μF. What is the duration? _____

☐ c. What is the status of LED D_1? _____ Of LED D_2? _____. Is the Q output LOW or HIGH? _____ Is the Q output LOW or HIGH? _____

TABLE 13.2

A_1	A_2	B	Q	$\overline{Q}$
↓	1	1	⎍	⎍
1	↓	1	⎍	⎍
0	x	↑	⎍	⎍
x	0	↑	⎍	⎍

☐   d. Momentarily press S_1, causing input $\overline{A}_1$ to transition from HIGH to LOW. Do both LEDs come on? _____ If so, why? _____

_____

Repeat the procedure and measure the pulse duration. What is the duration? _____

☐   e. Momentarily press S_2, causing input $\overline{A}_2$ to transition from HIGH to LOW. Do both LEDs come on? _____ If so, for how long?

_____

☐   3. To trigger the monostable circuit on a LOW-to-HIGH transition (positive edge trigger), input B is used. At the same time, either input $\overline{A}_1$ or $\overline{A}_2$ must be held LOW.

     ☐   a. Press and hold switch S_3 (input B) closed. Now press and hold switch S_1 (input $\overline{A}_1$) closed. Release S_3, causing a LOW-to-HIGH transition at input B. You can now release S_1. Do the LEDs come on?

_____. If so, for how long? _____

     ☐   b. Press and hold S_3 closed again. Press and hold S_2 closed. Release S_3. You can also release S_2. Do the LEDs come on? _____ If so, for how long? _____

☐   4. The 74121 monostable multivibrator is excellent for producing an extremely short pulse duration. To test this, replace the 470-μF capacitor, C_1, with a 0.1-μF capacitor.

     ☐   a. Calculate the pulse width for the one-shot circuit with an R_1 or 18 kΩ and a C_1 of 0.1 μF. What is the time? _____

     ☐   b. Using your dual-channel oscilloscope, connect one channel to the Q output. Press and release S_1. As you do, you should observe a brief pulse on the oscilloscope display. How wide is the pulse? _____ Repeat this procedure as often as necessary to obtain a good reading. Are your measured results consistent with the calculated pulse width?

_____

Part 3: Circuit Applications

A Light-Triggered Monostable Multivibrator: Dark/Light-Activated Timer Circuit

☐   1. A one-shot circuit that will output a pulse when a shadow is cast over one photoresistor, or when a light shines on another photoresistor, is shown in Figure 13.4. With jumper J_1 connected to point A, the circuit is "dark-activated." With light shining on photoresistor R_2, its resistance is LOW (around 500 Ω). The trigger input is thus held HIGH, R_3 providing the greater voltage drop in the R_2–R_3 voltage divider. When a shadow is cast over R_2, its resistance shoots up, to perhaps 500,000 Ω. Now the trigger input is pulled LOW, the voltage drop across R_3 being close to zero. As a result, the circuit outputs a pulse, the duration of which is determined by the values of R_1 and C_1.

The circuit is "light-activated" when J_1 is placed in position B. Explain how the light-activated circuit works: _____

_____

_____

_____

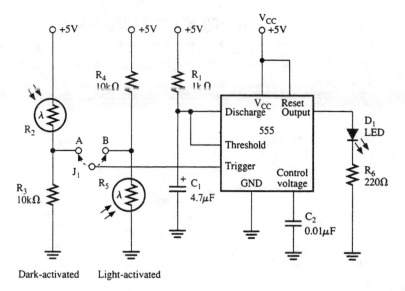

Dark-activated   Light-activated

FIGURE 13.4

☐ a. Referring to Figure 13.4, record missing IC pin numbers. Construct the circuit of Figure 13.4. Apply power. With an R_1 of 1 MΩ and a C_1 of 4.7 μF, what is the output pulse width? _____

☐ b. Place J_1 in position A while a light shines on R_2. Create a shadow across R_2. What happens to the LED? _____

_____

Repeat this procedure a few times.

☐ c. Place J_1 in position B while R_5 is in shadow. Shine a light on R_5. What happens to the LED? _____

_____

SUMMARY

In this experiment you assembled two monostable multivibrators, one using a 555, the other a 74121 IC. With the former, you varied the pulse width and examined reset and retrigger options. Using the 74121, you explored positive and negative edge triggering, the use of complementary outputs, and the creation of an extremely short pulse duration. Finally, you built and analyzed a Dark/Light-Activated Timer Circuit.

REVIEW QUESTIONS

1. A 555 monostable multivibrator has an R/C circuit with an R of 1.2 MΩ and a C of 150 μF. What is the output pulse width? _____

2. A 74121 monostable multivibrator has an R/C circuit with an R of 12 kΩ and a C of 0.001 μF. What is the output pulse width? _____

3. In Figure 13.3, if both LEDs are off, the Q output is _____; the $\overline{Q}$ output is _____.

4. The photoresistors used in the circuit of Figure 13.4 have a LOW resistance in the _____ and a HIGH resistance in the _____.

Writing Skills Assignment

In as few paragraphs as possible, explain how the Dark/Light-Activated Timer Circuit works. You might first discuss what the circuit does and then discuss how it does it.

14   S-R Flip-Flops

OBJECTIVES

After completing this experiment, you will be able to

- Assemble and then verify operation of a NAND gate active-LOW S-R flip-flop
- Assemble and then verify operation of a NOR gate active-HIGH S-R flip-flop
- Assemble and then verify operation of a clocked S-R flip-flop
- Build and analyze an S-R Flip-Flop Bounceless Switch Circuit

REFERENCE READING

Review Ronald Reis, *Digital Electronics Through Project Analysis*, Chapter 10, Section 10.1.

EQUIPMENT & MATERIALS NEEDED

Equipment

☐ 1 5-V power supply
☐ 1 solderless circuit board

Materials

☐ 1 7400 quad two-input NAND gate
☐ 1 7408 quad two-input AND gate
☐ 1 7432 quad two-input NOR gate
☐ 1 7490 decade counter
☐ 4 LEDs (red)
☐ 2 LEDs (green)
☐ 4 220-Ω resistors
☐ 3 1-kΩ resistors
☐ 1 N.O. push-button switch
☐ 1 package of jumper wires

BACKGROUND INFORMATION

The S-R (SET-RESET) flip-flop is the key element in sequential logic; from it all other flip-flops (D, T, and J-K) are derived.

The S-R flip-flop is set by forcing its Q output HIGH; it is reset by forcing its Q output LOW. There are two kinds of S-R flip-flops: active-HIGH and active-LOW (Figure

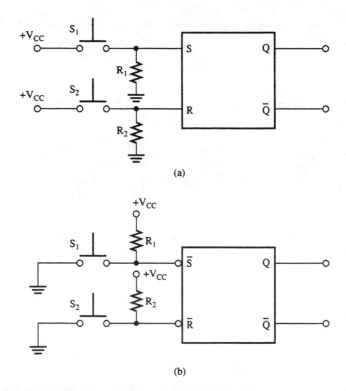

FIGURE 14.1 S-R flip-flops: (a) active-HIGH and (b) active-LOW.

14.1). With the former, it takes a HIGH pulse on the set (S) input to bring Q HIGH, while a HIGH on the reset (R) input brings Q LOW. For an active-LOW S-R flip-flop, a LOW pulse on the set input brings Q HIGH; a LOW on the reset input brings Q LOW.

In this experiment you will begin by building and testing an active-LOW and an active-HIGH S-R flip-flop using two NAND and two NOR gates, respectively. Then you will build and test a clocked S-R flip-flop containing two AND and two NOR gates. Finally, you will once again build and analyze a bounceless switch. This one, however, the S-R Flip-Flop Bounceless Switch Circuit, contains a flip-flop element.

PROCEDURE

Part 1: Circuit Fundamentals

Active-LOW and Active-HIGH S-R Flip-Flops

☐ 1. The schematic for an active-LOW S-R flip-flop using two two-input 7400 TTL NAND gates is shown in Figure 14.2. The partially completed truth table is presented as Table 14.1. As seen by referring to the table, in step 1 the S input is brought LOW; in step 2 it is returned to a HIGH. In step 3, the R input is brought LOW; in step 4 it is returned to HIGH. In step 5 both the S and R inputs are grounded, a forbidden condition.

    ☐ a. Referring to Figure 14.2, record missing IC pin numbers. Construct the circuit of Figure 14.2.

    ☐ b. Predict the output conditions for each step and record your results in the space provided in Table 14.1.

    ☐ c. Apply power. Set up the input conditions in each step, recording your measured output results in the space provided in Table 14.1.

    ☐ d. Are your measured results consistent with your predicted results?

       • In step 1, when S is brought LOW, does Q go HIGH and $\overline{Q}$ go LOW?

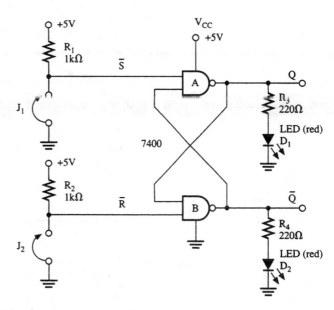

FIGURE 14.2

TABLE 14.1

Step	S	R	Predicted output Q	Predicted output $\overline{Q}$	Measured output Q	Measured output $\overline{Q}$
1	0	1				
2	1	1				
3	1	0				
4	1	1				
5	0	0				

- In step 2, when S is returned to a HIGH, do the Q and $\overline{Q}$ outputs remain the same? _____
- In step 3, when R is brought LOW, does Q go LOW and $\overline{Q}$ go HIGH? _____
- In step 4, when R is returned to a HIGH, do the Q and $\overline{Q}$ outputs remain the same? _____
- In step 5, when both S and R are brought LOW, are outputs Q and $\overline{Q}$ the same? That is, are both LOW or both HIGH? _____

☐ e. Summarize your results for the active-LOW flip-flop: _____
_____
_____

☐ 2. The schematic for an active-HIGH S-R flip-flop using two two-input 7402 TTL NOR gates is shown in Figure 14.3. The partially completed truth table is presented as Table 14.2. As seen by referring to the table, in step 1 the S input is brought HIGH; in step 2 it is returned to a LOW. In step 3, the R input is brought HIGH; in step 4 it is returned to a LOW. In step 5 both the S and R inputs are placed HIGH, a forbidden condition.

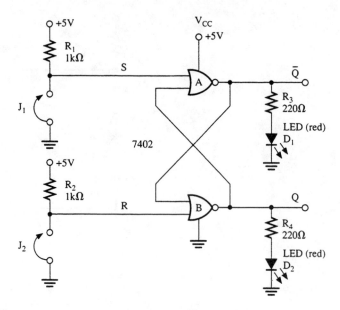

FIGURE 14.3

☐ a. Referring to Figure 14.3, record missing IC pin numbers. Construct the circuit of Figure 14.3.

☐ b. Predict the output conditions for each step and record your results in the space provided in Table 14.2.

☐ c. Apply power. Set up the input conditions in each step, recording your measured output results in the space provided in Table 14.2.

☐ d. Are your measured results consistent with your predicted results? _____

• In step 1, when S is brought HIGH, does Q go HIGH and $\overline{Q}$ go LOW? _____

• In step 2, when S is returned to a LOW, do the Q and $\overline{Q}$ outputs remain the same? _____

• In step 3, when R is brought HIGH, does Q go LOW and $\overline{Q}$ go HIGH? _____

• In step 4, when R is returned to a LOW, do the Q and $\overline{Q}$ outputs remain the same? _____

• In step 5, when both S and R are brought HIGH, are outputs Q and $\overline{Q}$ the same? That is, are both LOW or both HIGH? _____

TABLE 14.2

Step	S	R	Predicted output		Measured output	
			Q	$\overline{Q}$	Q	$\overline{Q}$
1	1	0				
2	0	0				
3	0	1				
4	1	0				
5	1	1				

☐ e. Summarize your results for the active-HIGH flip-flop: _____

_____

_____

Part 2: Further Investigation

The Clocked S-R Flip-Flop

☐ 1. The versatility of a standard S-R flip-flop can be greatly increased by providing gated inputs. By so doing, we can control when the set and reset signals are applied to the flip-flop. Such a clocked S-R flip-flop is shown in Figure 14.4.

When the clock input is LOW, both AND gates are disabled; thus each output can only be LOW. With LOWs on the S and R inputs to the flip-flop, its Q and $\overline{Q}$ outputs remain in their last states.

Even if the signals on the set and reset inputs change, as long as the clock input is LOW, the output of each AND gate remains LOW. Hence the flip-flop's outputs are unchanged.

When a positive clock pulse arrives, both AND gates are enabled for the duration of the clock pulse, one input of each gate is HIGH. The AND gates will now pass to their output whatever signal level is at their other input (set or reset). For instance, if during the clock pulse the set input is HIGH and the reset input LOW, S goes HIGH and R goes LOW. The flip-flop is set. On the other hand, if the set input is LOW and the reset input HIGH during the clock pulse, S goes LOW and R goes HIGH. The flip-flop is reset.

An active-HIGH clocked S-R flip-flop circuit using two AND and two OR gates is shown in Figure 14.5. The partially completed truth table is presented as Table 14.3.

☐ a. Referring to Figure 14.5, record missing IC pin numbers. Construct the circuit of Figure 14.5.

☐ b. Predict the output conditions for each step and record your results in the space provided in Table 14.3.

☐ c. Apply power. Set up the input conditions in each step and record your measured results in the space provided in Table 14.3. (*Note:* You will have to hold switch S_1 closed to maintain a HIGH on the clock input.)

☐ d. Are your measured results consistent with your predicted results?

_____

☐ e. Summarize your results for the clocked S-R flip-flop circuit: _____

_____

_____

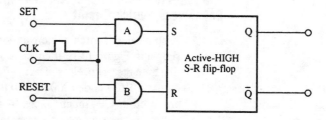

FIGURE 14.4

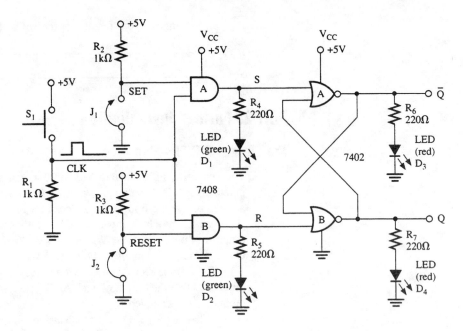

FIGURE 14.5

Step	CLK	SET	RESET	S	R	Predicted output		Measured output	
						Q	$\overline{Q}$	Q	$\overline{Q}$
1	0	0	0	0	0				
2	0	1	0	0	0				
3	0	0	1	0	0				
4	1	1	0	1	0				
5	0	1	0	0	0				
6	1	0	1	0	1				
7	0	0	1	0	0				
8	1	0	0	0	0				
9	1	1	1	1	1				

TABLE 14.3

Part 3: Circuit Applications

The Flip-Flop as a Bounceless Switch: S-R Flip-Flop
Bounceless Switch Circuit

☐ 1. A bounceless switch demonstration circuit using an active-LOW S-R flip-flop is shown in Figure 14.6. It consists of the bounceless switch at the left and, to help illustrate the bounceless effect, a 7490 decade counter on the right. The 7490 will output a BCD code (0000–1001) on LEDs D_1–D_4 when negative-going pulses arrive on its $\overline{CP_0}$ input (Table 14.4). If jumper J_2 is placed in position C, those pulses are generated by the pressing of push-button switch S_1. They are likely to consist of noisy, "dirty" spikes. If J_2 is in position D, pulses come from the

118

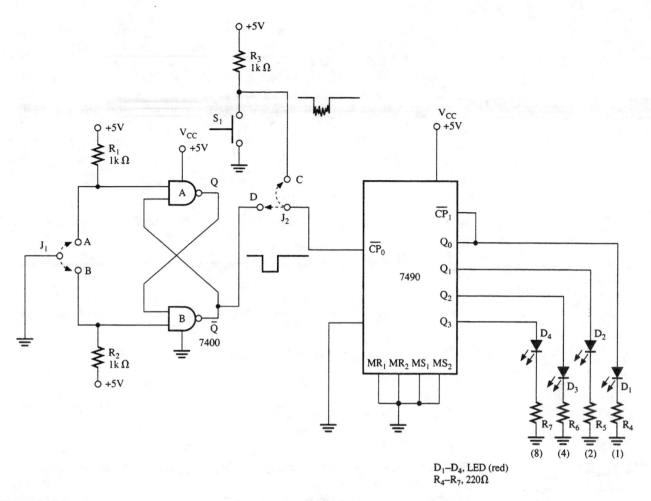

FIGURE 14.6

TABLE 14.4

Pulse 7400 $\overline{CP_0}$	D_4	D_3	D_2	D_1
	0	0	0	0
⊓	0	0	0	1
⊓	0	0	1	0
⊓	0	0	1	1
⊓	0	1	0	0
⊓	0	1	0	1
⊓	0	1	1	0
⊓	0	1	1	1
⊓	1	0	0	0
⊓	1	0	0	1
⊓	0	0	0	0

$\overline{Q}$ output of the S-R flip-flop. Now every time J_1 is moved from position B to position A and then back to position B, a negative, bounceless, glitch-free pulse appears at the input to the 7490 IC.

☐ a. Referring to Figure 14.6, record missing IC pin numbers. Construct the circuit of Figure 14.6. Apply power.

☐ b. Place J_2 in position C. (At this point it doesn't matter what you do with J_1.) Press and release S_1 a few times. As you do, note the count on LEDs D_1–D_4. Does the count advance in a smooth predictable sequence, as shown in Table 14.4, or is the count erratic, jumping all around?

_____ Repeat this procedure a few times.

☐ c. Now place J_2 in position D and J_1 in position B. Remove power from the circuit, then reapply. What is the status of LEDs D_1–D_4? _____

_____

☐ d. Place J_1 in position A and then back to position B, generating a negative pulse at output $\overline{Q}$. Does LED D_1 come on? _____

☐ e. Repeat step 1d to cause the binary count to proceed from 0000 to 1001. Does the count progress in a smooth, predictable, glitch-free manner?

_____

☐ f. Summarize the difference between what happens when J_2 is in position C compared with position D: _____

_____

_____

SUMMARY

This has been the first of three experiments dealing specifically with flip-flops. You began by investigating the simplest type, the active-LOW and active-HIGH S-R flip-flop. You then "progressed" to the clocked S-R flip-flop, which allows you to control when the input conditions are to affect the output. Finally, you built and then analyzed operation of an S-R Flip-Flop Bounceless Switch Circuit.

REVIEW QUESTIONS

1. With regard to Figure 14.2, explain why, when going from step 1 to step 2 in Table 14.1, no change occurs at the outputs. _____

_____

_____

2. With regard to Figure 14.3, explain why, when both the S and R inputs are HIGH, the Q and $\overline{Q}$ outputs are LOW. _____

_____

_____

3. Redesign the circuit of Figure 14.5 to operate with a negative clock pulse.

4. Referring to Figure 14.6, if point D of J_2 is connected to the Q instead of the $\overline{Q}$ output, will the circuit still work? _____ Explain. _____

_____

_____

Writing Skills Assignment

In as few paragraphs as possible, explain how the S-R Flip-Flop Bounceless Switch Circuit works. You might first discuss what the circuit does and then discuss how it does it.

15 D Flip-Flops

OBJECTIVES

After completing this experiment, you will be able to

- Construct and then verify operation of a D flip-flop using five NAND gates
- Analyze the operation of a TTL 7474 dual positive edge-triggered D flip-flop with preset, clear, and complementary outputs
- Analyze the operation of a TTL 7475 quad latch containing four D flip-flops
- Build and analyze a Latch Demonstration/Game Circuit

REFERENCE READING

Review Ronald Reis, *Digital Electronics Through Project Analysis*, Chapter 10, Section 10.2.

EQUIPMENT & MATERIALS NEEDED

Equipment

☐ 1 5-V power supply
☐ 1 solderless circuit board

Materials

☐ 1 7400 quad two-input NAND gate
☐ 1 7447 BCD-to-seven-segment decoder/driver
☐ 1 7474 dual positive edge-triggered D flip-flop
☐ 1 7475 quad latch
☐ 1 7490 decade counter
☐ 4 LEDs (red)
☐ 1 LED (green)
☐ 1 seven-segment common-anode LED display
☐ 7 220-Ω resistors
☐ 2 1-kΩ resistors
☐ 2 N.O. push-button switches
☐ 1 8-position DIP switch
☐ 1 package of jumper wires

FIGURE 15.1

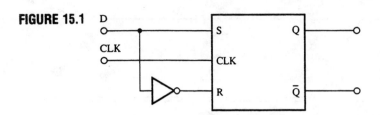

BACKGROUND INFORMATION

With the basic S-R flip-flop as a nucleus, the versatile and widely used D, or data, flip-flop can be created. Used as a latch or for constructing storage and shift registers, the D flip-flop has a clock input but only one data input. The basic logic symbol is shown in Figure 15.1. Note the inverter connected between the S and R inputs. Thus when the D input is HIGH, S is HIGH but R is LOW. When the D input is LOW, the reverse is true: S is LOW but R is HIGH. Therefore S and R are always at opposite states.

The basic D flip-flop operates by transferring the voltage level on the D input to the Q output when the clock is HIGH. When the clock is LOW, the Q and $\overline{Q}$ outputs remain in their last state.

In this experiment you will begin by constructing, then analyzing, a simple D flip-flop composed of five two-input TTL NAND gates. Next you will examine and test the TTL 7474 dual positive edge-triggered D flip-flop with preset, clear, and complementary outputs. You will also investigate the TTL 7475 quad latch containing four D flip-flops. Finally, you will build and analyze a Latch Demonstration/Game Circuit incorporating the 7475 IC.

PROCEDURE

Part 1: Circuit Fundamentals

D Flip-Flop Operation

☐ 1. One way to construct a D flip-flop is shown in Figure 15.2. When the clock is LOW, both AND gates are disabled, LOWs appear on the S and R inputs, and the flip-flop's outputs remain in their last state. If the clock is HIGH and D is HIGH, gate A is enabled while gate B is still disabled. Gate A outputs a HIGH and the flip-flop sets. If D is LOW while the clock is HIGH, gate A is disabled but gate B is enabled. Gate B outputs a HIGH and the flip-flop resets.

If we substitute NAND gates for AND gates and use NAND gates to create the S-R flip-flop, we have the positive pulse-triggered, five-NAND-gate D flip-flop shown in Figure 15.3. Note that if we tie the two inputs of gate A together, an inverter, or NOT gate, is created.

   ☐ a. Referring to Figure 15.3, record missing IC pin numbers. Construct the circuit of Figure 15.3. Do not apply power yet.

   ☐ b. Set up the input conditions shown in step 1 of Table 15.1. Apply power. Record the output conditions in the space provided. Is the flip-flop set or

   reset? _____

   ☐ c. Set up the input conditions shown in step 2 of Table 15.1. Record the output conditions in the space provided. Are the outputs the same as those

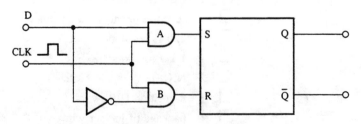

FIGURE 15.2

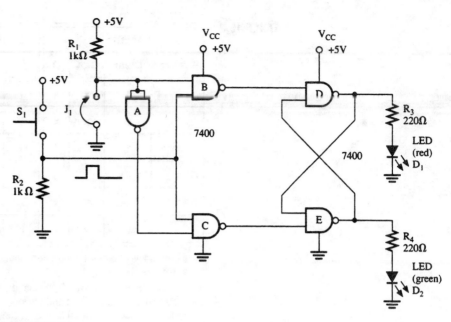

FIGURE 15.3

TABLE 15.1

Step	D	CLK	Q	$\overline{Q}$
1	0	0		
2	1	0		
3	1	⎍		
4	0	⎍		

in step 1? _____ Why is this so? _____

_____

☐ d. Set up the input conditions shown in step 3 of Table 15.1. (*Note:* To provide a positive pulse trigger, momentarily press switch S_1.) Record the output conditions in the space provided. Does the flip-flop set?

_____ Does the Q output follow the D input? _____

☐ e. Set up the input conditions shown in step 4 of Table 15.1. Record the output conditions in the space provided. Does the flip-flop reset?

_____ Does the Q output follow the D input? _____

☐ f. Summarize your findings with regard to the D flip-flop: _____

_____

_____

_____

Part 2: Further Investigation

A D Flip-Flop IC

☐ 1. You do not have to build D flip-flops from discrete gates; many are available in TTL and CMOS IC packages. One such IC is the TTL 7474 dual positive edge-triggered D flip-flop with preset, clear, and complementary outputs, the pin

TABLE 15.2

Inputs				Outputs	
Preset	Clear	CLK	D	Q	$\overline{Q}$
0	1	x	x	1	0
1	0	x	x	0	1
1	1	↑	1	1	0
1	1	↑	0	0	1
1	1	0	x	Q_0*	$\overline{Q}_0$*
0	0	x	x	$1^\dagger$	$1^\dagger$

*The output level before the indicated input conditions were established.

†This configuration is nonstable; that is, it will not persist when either the preset and/or clear inputs return to their inactive (HIGH) level.

configuration of which is shown in Appendix A (Figure A.9). The IC's truth table is presented as Table 15.2.

The device contains two positive edge-triggered D flip-flops with complementary outputs: Q and $\overline{Q}$. The data at the D input goes to the Q output whenever the clock input changes from a LOW to a HIGH level. If the circuit is not clocked, changes on the D input are not passed on to the Q output.

If D is HIGH, on the positive edge of the clock pulse, Q goes HIGH and $\overline{Q}$ goes LOW. If D is LOW, on the positive edge of the clock pulse, Q goes LOW and $\overline{Q}$ goes HIGH.

Note that information on the D input can be changed at any time. Yet it is only the value at the instant of the positive clock edge that counts; it is this data that is entered into the flip-flop.

Under normal operation, the preset (set) and clear (reset) inputs are tied HIGH. However, if the preset input is momentarily grounded, the flip-flop immediately goes into the set state with Q high and $\overline{Q}$ LOW. If the clear input is momentarily grounded, the flip-flop immediately goes into the reset state with Q LOW and $\overline{Q}$ HIGH. Preset and clear should never be grounded simultaneously or a disallowed state will exist.

☐ a. Referring to Figure 15.4, record missing IC pin numbers. Construct the 7474 test circuit shown in Figure 15.4. Do not apply power yet.

☐ b. Using Table 15.3, you are asked to confirm the 7474 truth table. In section A, steps 1 and 2, you will examine what happens at the Q and $\overline{Q}$ outputs when power is first applied to the IC. Set up the input conditions in step 1 and apply power to the circuit. Record the output results in the space provided. Remove power from the circuit. Set up the input conditions in step 2 and apply power to the circuit. Record the

output results in the space provided. Summarize your results: _____

_____

_____

☐ c. In section B, steps 3–8, you will discover how the preset and clear inputs override all other inputs; that is, no matter what is happening on the D or clock inputs, bringing preset or clear LOW will override them. Set up the input conditions in steps 3–8 and record the output results in the space provided. Are your results consistent with the chip's truth table?

_____

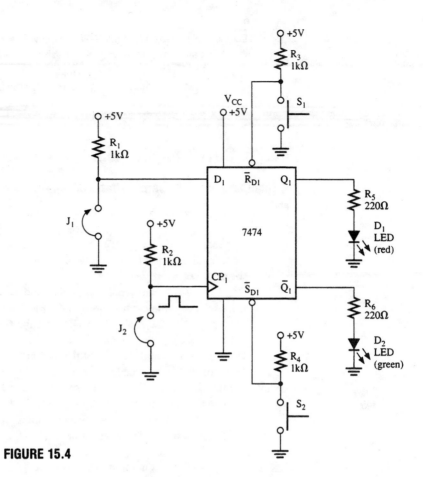

FIGURE 15.4

	Step	Preset	Clear	CLK	D	Q	$\overline{Q}$
	1	1	1	0	1		
Section A	2	1	1	0	0		
	3	⊓⊔	1	0	1		
	4	1	⊓⊔	0	1		
	5	⊓⊔	1	0	0		
Section B	6	1	⊓⊔	0	0		
	7	⊓⊔	1	1	1		
	8	1	⊓⊔	1	0		
	9	1	1	0	0		
	10	1	1	0	1		
	11	1	1	↑	1		
Section C	12	1	1	0	1		
	13	1	1	0	0		
	14	1	1	↑	0		

TABLE 15.3

d. In section C, steps 9–14, you will see how the data on the D input is transferred to the Q output only during a positive edge trigger. Set up the input conditions in steps 9–14 and record the output results in the space provided. Are your results consistent with the chip's truth table?

_____

Part 3: Circuit Applications

A Quad Latch: Latch Demonstration/Game Circuit

☐ 1. Another type of D flip-flop is found in the TTL 7475 level-sensitive quad latch with complementary outputs, the pin configuration of which is shown in Appendix A (Figure A.10). This device is ideally suited for use as temporary storage for binary information between processing units and digital indicators, as we shall see shortly.

The 7475 package contains four memory elements that are controlled in pairs by an enable control. Information present at a D input is transferred to the Q output when the enable is HIGH, and the Q output will follow the data input as long as the enable remains HIGH. When the enable goes LOW, the data on the D input retained at the Q output the instant the transition occurs is retained at the Q output until the enable is permitted to go HIGH. In other words, with the enable HIGH, the IC is "transparent" to incoming data, the data proceeds through the chip, from D to Q, as though the IC were not there. When the enable goes LOW, whatever data is present at the D input is held on the Q output. The 7475 truth table is shown as Table 15.4.

☐ a. Referring to Figure 15.5, record missing IC pin numbers. Construct the 7475 test circuit shown in Figure 15.5. Apply power.

☐ b. Open jumper J_1, bringing both enable pins HIGH. Using switches S_1–S_4, set up various input conditions and note the Q outputs. Does each Q

output follow its D input? _____

☐ c. Note the last output condition. Now close J_1, bringing both enable pins LOW. Using S_1–S_4, change the input conditions. Does the Q output

follow its new D input? _____ Why is this so? _____

_____

☐ 2. A Latch Demonstration/Game Circuit, using the 7475 IC, is shown in Figure 15.6. It is essentially a single-digit decade counter. Pulses arriving on the $\overline{CP}_0$ input of the 7490 IC are counted, decoded by the 7447, and displayed on the seven-segment readout. If the input signal is at 1 Hz, the readout will advance from 0 through 9, and repeat, at a 1-s rate.

Note, of course, that a 7475 quad latch is placed between the 7490 decade counter and the 7447 decoder/driver. Yet as long as enable inputs (EN_{0-1}, EN_{2-3}) are held HIGH, the circuit acts as though the 7475 were not even there. But if switch S_1 is pressed, bringing the enable pins LOW, the binary count on the D

TABLE 15.4

Inputs		Outputs	
D	EN	Q	$\overline{Q}$
0	1	0	1
1	1	1	0
x	0	Q_0	$\overline{Q}_0$

FOR EACH LATCH

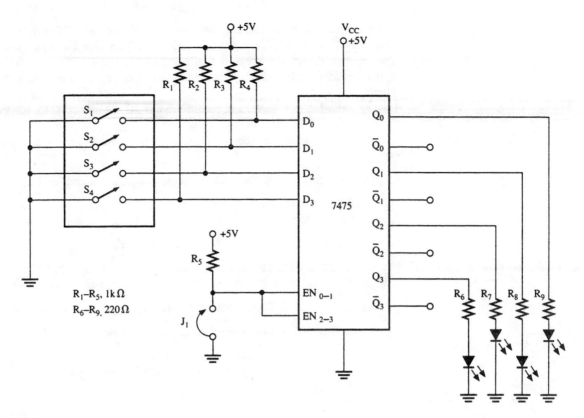

FIGURE 15.5

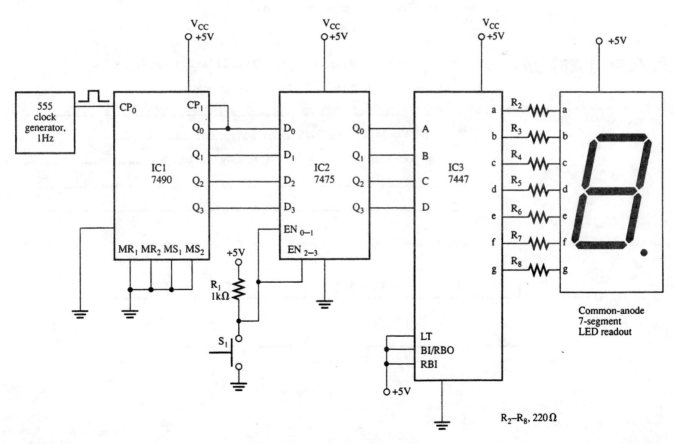

FIGURE 15.6

129

inputs to the 7475 the instant the switch is closed will "freeze" on the seven-segment readout. As long as S_1 is held closed, the *displayed* count does not change (even though the circuit is continuing to count at a 1-Hz rate.) When S_1 is released, the readout displays the binary count currently on the 7475 D_1–D_4 inputs.

☐   a. Referring to Figure 15.6, record missing IC pin numbers. Construct the circuit of Figure 15.6. Apply power.

☐   b. Watch the display count. Does it advance from 0 to 9 and repeat? _____

☐   c. Immediately after the readout displays a 7, press and hold S_1 closed. Does the 7 continue to be displayed? _____

☐   d. Set the 555 clock generator to 1 kHz. What number appears on the readout? _____ Why is this so? _____ _____

☐   e. Press and hold S_1 closed. What number appears on the readout? _____ Repeat this procedure a few times. Can you predict the number that will come up? _____ If not, why not? _____ _____

You now have a "digital guessing game." Have some fun!

SUMMARY

In this experiment you began by constructing, then testing, a D flip-flop from discrete gates. Next you analyzed the operation of a dual positive edge-triggered D flip-flop, the TTL 7474. You moved on to investigate the TTL 7475 quad latch containing four D flip-flops. Finally, you built and then analyzed the operation of a Latch Demonstration/Game Circuit that incorporates the 7475 IC.

REVIEW QUESTIONS

1. Complete the timing diagram for the 7475 quad latch, using Figure 15.7.

2. When found in a truth table, what do Q and $\overline{Q}$ mean? _____ _____

3. Which IC, the 7474 or the 7475, is level-triggered? _____

4. Why are both enable pins of a 7475 IC usually tied together? _____ _____ _____

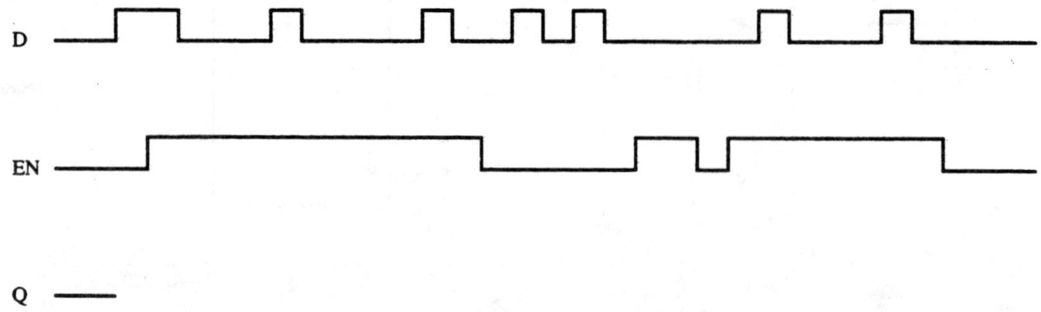

FIGURE 15.7

Writing Skills Assignment

In as few paragraphs as possible, explain how the Latch Demonstration/Game Circuit works. You might first discuss what the circuit does and then discuss how it does it.

16 J-K Flip-Flops

OBJECTIVES

After completing this experiment, you will be able to

- Verify operation of the TTL 7473 J-K flip-flop
- Verify operation of a J-K flip-flop as a frequency divider and binary counter
- Build and analyze a Delay Roulette Wheel Circuit

REFERENCE READING

Review Ronald Reis, *Digital Electronics Through Project Analysis*, Chapter 10, Section 10.2.

EQUIPMENT & MATERIALS NEEDED

Equipment

☐ 1   5-V power supply
☐ 1   555 clock generator
☐ 1   logic pulser
☐ 1   dual-channel oscilloscope
☐ 1   solderless circuit board

Materials

☐   2   7473 dual master-slave J-K flip-flops
☐   1   74132 quad two-input Schmitt trigger NAND gate
☐   1   74154 4-line-to-16 line decoder/demultiplexer
☐ 20   LEDs (red)
☐   1   LED (green)
☐   1   1000-μF capacitor
☐   4   220-Ω resistors
☐   3   1-kΩ resistors
☐   1   N.O. push-button switch
☐   1   package of jumper wires

FIGURE 16.1

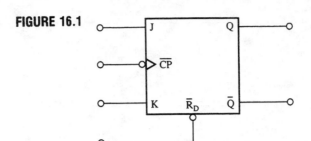

BACKGROUND INFORMATION

The J-K flip-flop, the most widely used of all flip-flops, can be wired to perform as an S-R, a D, and a T (toggle) flip-flop. In the latter mode, it acts as a frequency divider or binary counter. The block diagram for a negative edge triggered J-K flip-flop with active-LOW clear and complementary outputs is shown in Figure 16.1. The truth table is presented as Table 16.1.

Looking at Table 16.1, we note, in step 1, that as long as the clear input ($\overline{R}_D$) is LOW, it doesn't matter what is happening on the $\overline{CP}$, J, and K inputs; Q is LOW and $\overline{Q}$ is HIGH. In steps 2–5, the clear input is held HIGH and the flip-flop will respond to an input clock pulse. In step 2, the clock goes HIGH then LOW, but since the J and K inputs are LOW, the outputs do not change from their previous state. In step 3, when the clock goes HIGH then LOW, if J is HIGH and K is LOW, the flip-flop sets, with Q going HIGH and $\overline{Q}$ going LOW. In step 4, the reverse is true. When the clock goes HIGH then LOW, if J is LOW and K is HIGH, the flip-flop resets, with Q going LOW and $\overline{Q}$ going HIGH. Finally, and most interesting, note carefully step 5. If the J and K inputs are held HIGH and the clock goes HIGH then LOW, the flip-flop *toggles,* that is, its outputs change states (and will do so with every additional HIGH-to-LOW transition of an incoming clock pulse). In the toggle mode, as we shall see shortly, the flip-flop acts as a frequency divider and binary counter.

In this experiment you will begin by investigating the TTL 7473 dual master-slave J-K flip-flop with clear and complementary outputs. Next you will see how J-K flip-flops can operate as frequency dividers and binary counters. In Part 3 you will build and analyze operation of a Delay Roulette Wheel Circuit that illustrates the binary counting function of this most versatile of all flip-flops.

PROCEDURE

Part 1: Circuit Fundamentals

A Dual J-K Flip-Flop

☐ 1. A test circuit for the 7473 J-K flip-flop, using one of its two internal flip-flops, is shown in Figure 16.2. The IC's truth table, Table 16.2, is identical to the one presented in Table 16.1.

Note that the 7473 is a master-slave flip-flop. The J and K data is processed by the flip-flop only after a *complete* clock pulse. While the clock is LOW, the

TABLE 16.1

Step	Inputs				Outputs	
	$\overline{R}_D$	$\overline{CP}$	J	K	Q	$\overline{Q}$
1	0	x	x	x	0	1
2	1	⎍	0	0	0	1
3	1	⎍	1	0	1	0
4	1	⎍	0	1	0	1
5	1	⎍	1	1	Toggle	

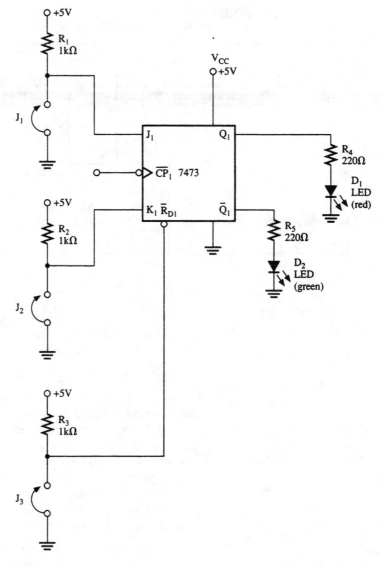

FIGURE 16.2

slave is isolated from the master. On the *positive* transition of the clock, the data from the J and K inputs is transferred to the master. While the clock remains HIGH, the J and K inputs are disabled. On the *negative* transition of the clock, the data from the master is transferred to the slave (and the Q and $\overline{Q}$ outputs).

☐ a. Referring to Figure 16.2, record missing IC pin numbers. Construct the circuit shown in Figure 16.2. Do not apply power.

TABLE 16.2

Step	Inputs				Outputs	
	$\overline{R}_D$	$\overline{CP}$	J	K	Q	$\overline{Q}$
1	0	x	x	x		
2	1	⊓	0	0		
3	1	⊓	1	0		
4	1	⊓	0	1		
5	1	⊓	1	1		

FIGURE 16.3

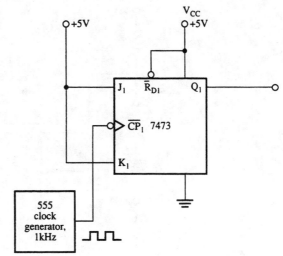

□ b. Via jumper J_3, tie the clear $\overline{R}_{D1}$ input LOW (step 1 of Table 16.2). You can leave the $\overline{CP}_1$, J, and K inputs HIGH or floating. Apply power to the circuit. Record the indicated results in the space provided in Table 16.2.

□ c. Set up the input conditions shown in steps 2–5 of Table 16.2. Use your logic pulser to apply a positive pulse trigger. Record the indicated results in the space provided.

□ 2. With the J and K inputs tied HIGH, the J-K flip-flop toggles with every clock pulse. Thus it becomes a divide-by-2 frequency divider.

□ a. Referring to Figure 16.3, record missing IC pin numbers. Construct the circuit of Figure 16.3. Apply power.

□ b. Using your dual-channel oscilloscope, connect channel 1 to the clock input of the J-K flip-flop and channel 2 to the Q_1 output. Obtain a stable wave pattern on the oscilloscope screen. Draw the waveforms in Plot 16.1.

□ c. Using the oscilloscope, measure the input and output frequencies. What is the input (clock) frequency? _____. What is the output frequency at Q_1? _____ Is the flip-flop functioning as a divide-by-2 frequency divider? _____

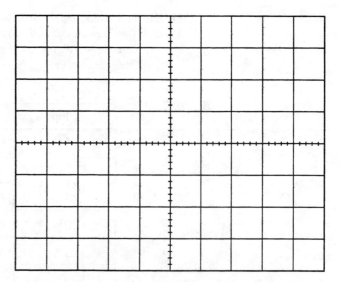

PLOT 16.1

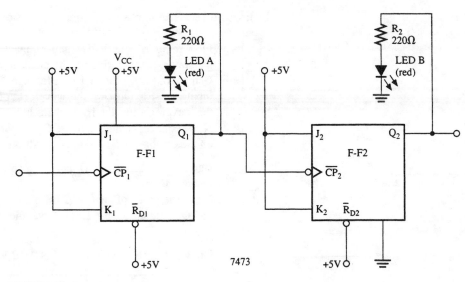

FIGURE 16.4

Part 2: Further Investigation

A Frequency Divider and Binary Counter

☐ 1. By cascading two-J-K flip-flops, you can create a divide-by-4 frequency divider and quad binary counter. The circuit is shown in Figure 16.4. The truth table is presented as Table 16.3.

   ☐ a. Referring to Figure 16.4, record missing IC pin numbers. Construct the circuit of Figure 16.4. Apply power.

   ☐ b. Using your logic pulser, "pulse" $\overline{CP}_1$ until both LEDs are off. Next, apply a pulse to $\overline{CP}_1$ and note the status of both LEDs (off or on). Record the results in the space provided in Table 16.3. Continue to apply clock pulses, each time recording the status of LEDs A and B in the space shown.

   ☐ c. Based on your observations, is F-F1 dividing the input pulses by 2? _____ Is F-F2 dividing the input pulses by 4? _____

   ☐ d. Note the binary count produced in Table 16.3. Does the circuit count from 0 to 4 in binary and then repeat? _____ Have you created a quad binary counter? _____

TABLE 16.3

$\overline{CP}_1$	Q_2 B	Q_1 A	LED B	LED A
⊓	0	0	OFF	OFF
⊓	0	1		
⊓	1	0		
⊓	1	1		
⊓	0	0		
⊓	0	1		
⊓	1	0		
⊓	1	1		

137

2. It would be useful to observe frequency division in the circuit of Figure 16.4 on an oscilloscope.
   a. First remove both LEDs and connect the output of your 555 clock generator, set to 1 kHz, to the clock input of F-F1. Apply power.
   b. Using your dual-channel oscilloscope, connect channel 1 to the clock input of F-F1. Observe the waveform on the oscilloscope screen. Draw the waveform in Plot 16.2 where indicated.
   c. Next, connect channel 2 to the Q_1 output of F-F1. Observe the waveform on the oscilloscope screen. Is it half the frequency of the incoming clock

   signal? _____ Draw the second waveform in Plot 16.2 where indicated.
   d. Move the probe of channel 2 from the Q output of F-F1 and connect it to the Q_2 output of F-F2. Observe the waveform on the screen. Is it

   one-fourth the frequency of the incoming clock signal? _____ Draw the third waveform in Plot 16.2 where indicated.

Clock input ⟶

Output of F-F1 ⟶

Output of F-F2 ⟶

PLOT 16.2

3. A divide-by-16 frequency divider and hexadecimal binary counter can be constructed by cascading four J-K flip-flops. The circuit, using two 7473 ICs, is shown in Figure 16.5. The truth table is presented as Table 16.4 (p. 140).
   a. Referring to Figure 16.5, record missing IC pin numbers. Construct the circuit of Figure 16.5. Apply power.
   b. Using your logic pulser, "pulse" CP_1 until all four LEDs are off. Next, apply a pulse to CP_1 of F-F1 and note the status of the LEDs (off or on). Record the results in the space provided in Table 16.4. Continue to apply clock pulses, each time recording the status of LEDs A, B, C, and D in the space shown.
   c. Note the binary count produced in Table 16.4. Is the circuit counting

   from 0 to 16 in binary? _____ Is the frequency at Q_1 of F-F1

   one-half that of the clock frequency? _____ At Q_2 of F-F2

   one-fourth that of the clock frequency? _____ At Q_1 of F-F3

   one-eighth that of the clock frequency? _____ And at Q_2 of F-F4

   one-sixteenth that of the clock frequency? _____ If so, why is

   this so? _____

   _____

   Note: Do not disassemble the circuit.

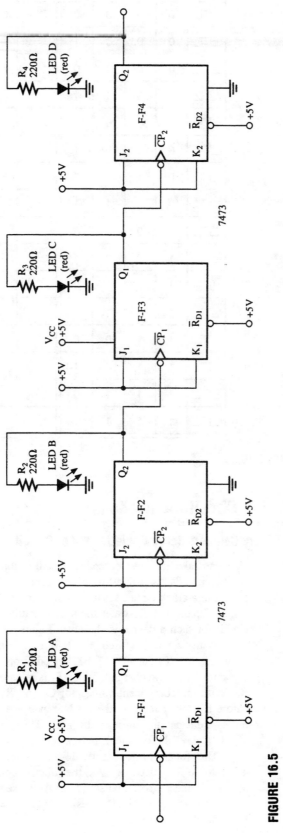

FIGURE 16.5

139

$\overline{CD_1}$	Q_4 (D)	Q_3 (C)	Q_2 (B)	Q_1 (A)	LED D	LED C	LED B	LED A
⎍	0	0	0	0	OFF	OFF	OFF	OFF
⎍	0	0	0	1				
⎍	0	0	1	0				
⎍	0	0	1	1				
⎍	0	1	0	0				
⎍	0	1	0	1				
⎍	0	1	1	0				
⎍	0	1	1	1				
⎍	1	0	0	0				
⎍	1	0	0	1				
⎍	1	0	1	0				
⎍	1	0	1	1				
⎍	1	1	0	0				
⎍	1	1	0	1				
⎍	1	1	1	0				
⎍	1	1	1	1				

TABLE 16.4

Part 3: Circuit Applications

A Binary Counter: Delay Roulette Wheel Circuit

☐ 1. You can take the four cascaded J-K flip-flops of Figure 16.5 and, with the addition of a 74154 4-line-to-16-line decoder, a Schmitt trigger two-input NAND gate, and a clock circuit set to approximately 10 Hz, construct a Delay Roulette Wheel Circuit that illustrates the binary counting function of a J-K flip-flop. The circuit for such a device is shown in Figure 16.6.

With switch S_1 closed, the NAND gate passes pulses from the 555 clock generator to $\overline{CP_1}$ input of F-F1. In addition, as long as S_1 remains closed, capacitor C_1 is held charged. When S_1 is released, C_1 discharges through resistor R_1. With the component values for C_1 and R_1 as indicated, it takes about 3 s before the lower input to the NAND gate goes LOW, preventing further clock pulses from passing through the gate. Thus, with the release of S_1, there is a slight "delay" before clock pulses cease to arrive on $\overline{CP_1}$ F-F1.

The four flip-flops provide the ABCD binary count for the 74154 4-line-to-16-line decoder. As the binary count on its four decoder inputs advances, one of 16 outputs goes LOW in sequence, lighting an LED in turn. Thus you have an "LED guessing game" with a built-in delay—a Delay Roulette Wheel Circuit.

☐ a. Referring to Figure 16.6, record the missing IC pin numbers. (Look up the 74154 pin numbers in Appendix A; see Figure A.23.) Construct the circuit of Figure 16.6. Apply power.

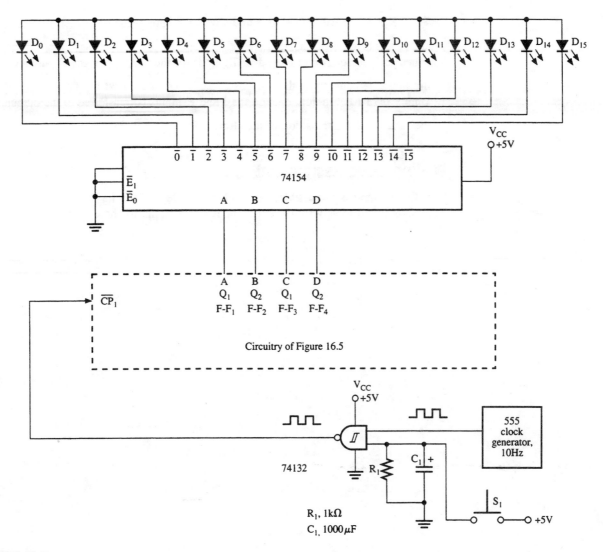

FIGURE 16.6

b. What is the status of LEDs A–D? _____ What is the status of LEDs D_0–D_{15}? _____ Press and hold S_1 closed. Do LEDs A–D "count" in binary? _____ Do LEDs D_0–D_{15} "progress" in sequence? _____ Release S_1. Do LEDs D_0–D_{15} continue to light in sequence and then "stop" approximately 3 s later? _____ Press, hold, and release S_1 a few times to repeatedly test the circuit.

SUMMARY

In this experiment you began by investigating the TTL 7473 J-K flip-flop. You saw how it operates as a frequency divider and as a binary counter. You cascaded flip-flops to produce a divide-by-16 and hexadecimal binary counter. You concluded by building and then analyzing operation of a Delay Roulette Wheel Circuit.

REVIEW QUESTIONS

1. The J-K flip-flop maintains the previous output condition with the application of a clock pulse if the clear ($\overline{R}_D$) input is _____, the J input is _____, and the K input is _____.

2. The J-K flip-flop will toggle its outputs with the application of a clock pulse if the J and K inputs are _____.

3. When you cascade two or more J-K flip-flops (of the 7473 type), the Q output of one flip-flop is fed directly to the _____ input of the next flip-flop.

4. With a 7473 J-K flip-flop, data on the J and K inputs is transferred to the outputs on the _____ transition of the clock pulse.

Writing Skills Assignment

In as few paragraphs as possible, explain how the Delay Roulette Wheel Circuit works. You might first discuss what the circuit does and then discuss how it does it.

17 Shift Registers

OBJECTIVES

After completing this experiment, you will be able to

- Verify operation of a 74194 shift register in the serial shift-right/shift-left modes
- Verify operation of a 74194 shift register in the parallel (broadside) load and inhibit clock modes
- Build and analyze a Programmable Sequence Generator Circuit using two 74194 shift registers

REFERENCE READING

Review Ronald Reis, *Digital Electronics Through Project Analysis,* Chapter 11, Section 11.1.

EQUIPMENT & MATERIALS NEEDED

Equipment

- ☐ 1 5-V power supply
- ☐ 1 logic pulser
- ☐ 1 555 clock generator
- ☐ 1 solderless circuit board

Materials

- ☐ 2 74194 4-bit bidirectional universal shift registers
- ☐ 8 LEDs (red)
- ☐ 2 0.1-μF capacitors
- ☐ 8 220-Ω resistors
- ☐ 9 1-kΩ resistors
- ☐ 1 N.C. push-button switch
- ☐ 1 8-position DIP switch (or equivalent)
- ☐ 1 package of jumper wires

BACKGROUND INFORMATION

A shift register is a circuit used to store information by shifting data from one stage (flip-flop) to another. As such, it can be used in sequencers and counters and to store, delay, and format information.

Shift registers are organized primarily on the basis of how their individual stages are accessed. There are four types: (1) serial-in/serial-out (SISO); (2) serial-in/parallel-out (SIPO); (3) parallel-in/serial-out (PISO); and (4) parallel-in/parallel-out (PIPO).

With the SISO shift register, data is input to the first stage and output from the last stage. With each clock pulse, data is moved one stage from left to right (in a shift-right register) or from right to left (in a shift-left register).

A SIPO shift register is also a SISO shift register except that, in addition, the output of every stage (flip-flop) is available. Data enters the first stage but is "sampled" at all stages as it moves from left to right (or right to left).

The PISO shift register is also an SISO, but it allows for the entering of all bits of a digital word at once and then clocks them out 1 bit at a time. Thus it is a parallel-to-serial converter.

Finally, a PIPO shift register allows for the entering and exiting of data in "broadside," or parallel. As such it acts as a conventional storage register.

A shift register that combines all of the above types, and is available in a 16-pin DIP package, is the TTL 74194 4-bit bidirectional universal shift register, the pin configuration of which is shown in Appendix A (Figure A.26). The 74194 features parallel inputs, parallel outputs, shift-right, shift-left, serial inputs, operating-mode-control inputs, and a direct overriding clear line.

In this experiment you will use the 74194 IC to examine the shift-right, shift-left, parallel load, and inhibit functions. You will conclude by building, then analyzing, a Programmable Sequence Generator Circuit that illustrates many universal shift register features.

PROCEDURE

Part 1: Circuit Fundamentals

Shift-Right/Shift-Left Functions

☐ 1. A test circuit for the 74194 4-bit shift register is shown in Figure 17.1. Four outputs are available: Q_0, Q_1, Q_2, and Q_3. There is the active-LOW clear input $\overline{MR}$, and the positive edge-triggered clock input, CP. S_0 and S_1 are the mode-control inputs. When S_0 is HIGH and S_1 is LOW, the IC is in the shift-right mode. When S_0 is LOW and S_1 is HIGH, the 74194 is in the shift-left mode. If both S_0 and S_1 are HIGH, the shift register is ready to do a parallel (broadside) load. Last, if both S_0 and S_1 are LOW, the internal flip-flops are inhibited—their outputs remain in their previous state.

DSR stands for shift-right data. If the IC is in the shift-right mode, data to be shifted is entered on this pin. DSL stands for shift-left data. If the 74194 is in the shift-left mode, data to be shifted is entered on this pin. Finally, we have the parallel inputs, labeled D_0, D_1, D_2, and D_3. Table 17.1 presents the partially completed truth table for the 74194.

☐ a. Referring to Figure 17.1, record missing IC pin numbers. Construct the circuit of Figure 17.1. Do not apply power yet.

☐ b. Set up the input conditions in step 1 of the truth table. Place the clear ($\overline{MR}$) input LOW. All other inputs can be in any condition, LOW or HIGH. Apply power to the circuit. Record the output conditions in the

space provided. What is the state of the outputs? _____
Leaving the clear input LOW, change the status of various other inputs.

Does the status of the outputs change? _____ You can conclude that with the clear input LOW, it doesn't matter what the other inputs are

doing, all outputs will remain _____.

☐ c. In steps 2–5 of the truth table you will examine how the shift register shifts-right the data on the DSR input with every positive edge-triggered clock pulse. Place the clear input HIGH, the S_0 input HIGH, and the S_1 input LOW. It doesn't matter what you do with the DSL input since it is used for shifting data to the left.

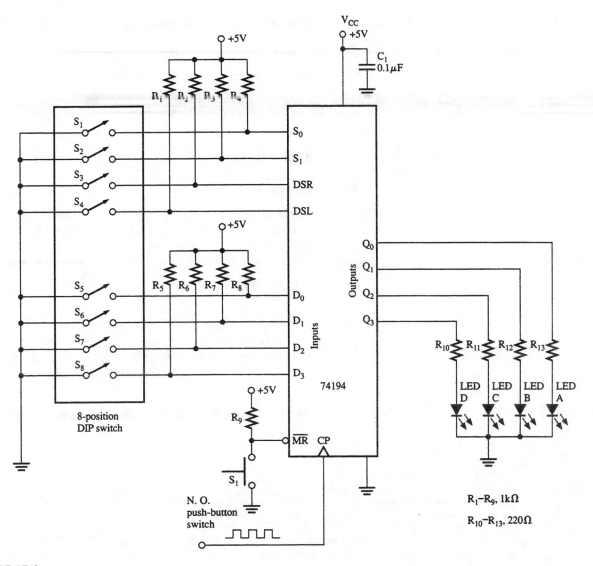

FIGURE 17.1

In step 2, enter a HIGH on the DSR input. Using your logic pulser, supply a positive edge trigger to the clock input. Record the output conditions in the space provided.

Proceed to steps 3, 4, and 5, entering the indicated data on the DSR input and then supplying the clock pulse. Record the results where indicated.

☐ d. Examine the outputs in steps 2–5 carefully. What is happening to the

input data with each clock pulse? _____

_____

How does the output data in step 5 compare with the input data in steps

2–5? _____

_____

☐ e. Bring the clear input LOW to clear the output data lines, step 6. Are any

LEDs on? _____ Why is this so? _____

_____

☐ f. To examine how the shift register shifts left, set up the input conditions in steps 7–10 as you did for steps 2–5. Record your outputs in the space provided. What are your conclusions with regard to the shift-left operation? _____

_____

_____

Note: Do not remove power from the circuit.

Part 2: Further Investigation

Parallel Load and Inhibit Functions

☐ 1. Parallel loading is accomplished by applying 4 bits of data to the D_0–D_3 inputs while S_0 and S_1 are HIGH. Each data bit is loaded into an associated flip-flop and appears at an output after the positive transition of the clock pulse.

    ☐ a. Clear the registers of your 74194 test circuit by momentarily bringing the clear input LOW. What is the status of the output LEDs? _____

_____

    ☐ b. Using switches S_5–S_8, place a binary 1010 on inputs D_0–D_3. Using your logic pulser, provide a positive edge-triggered clock pulse to the clock input. Record the status of the output LEDs in the space provided in Table 17.1 (steps 11 and 12). Does the input data shift to the output?

_____

☐ 2. Clocking can be inhibited by bringing S_0 and S_1 LOW. This is the "do-nothing" state.

Step		Inputs										Outputs			
		Mode			**Serial**		**Parallel**								
	$\overline{MR}$	S_1	S_0	CP	Left	Right	D_0	D_1	D_2	D_3		Q_0	Q_1	Q_2	Q_3
1	0	x	x	x	x	x	x	x	x	x					
2	1	0	1	↑	x	1	x	x	x	x					
3	1	0	1	↑	x	0	x	x	x	x					
4	1	0	1	↑	x	1	x	x	x	x					
5	1	0	1	↑	x	1	x	x	x	x					
6	0	x	x	x	x	x	x	x	x	x					
7	1	1	0	↑	1	x	x	x	x	x					
8	1	1	0	↑	0	x	x	x	x	x					
9	1	1	0	↑	1	x	x	x	x	x					
10	1	1	0	↑	1	x	x	x	x	x					
11	1	1	1	↑	x	x	0	1	0	1					
12	1	1	1	0	x	x	x	x	x	x					
13	1	0	0	x	x	x	x	x	x	x					

Note: "Shift right" brackets steps 2–5; "Shift left" brackets steps 7–10.

TABLE 17.1

146

a. Bring S_0 and S_1 LOW.

b. Place a binary 0101 on inputs D_0–D_3. Using your logic pulser, produce a positive edge-triggered clock pulse to the clock input. Record the status of the output LEDs in the space provided in Table 17.1 (step 13). Does the input data shift to the output? _____ Why is this so? _____

Part 3: Circuit Applications

Cascading Shift Registers: Programmable Sequence Generator Circuit

1. A shift register can be made to recirculate its own data by bringing the serial output around to the serial input. In that way a bit pattern appearing on the parallel outputs is repeatedly shifted to the right or left at a fixed rate.

   A Programmable Sequence Generator Circuit employing the recirculating principle, and thus illustrating parallel load and serial-right data shifting, is shown in Figure 17.2. It consists of a clock circuit, two cascaded 74194 shift registers, eight input switches, eight LEDs, and a few additional components. With this circuit, LEDs can be made to "chase each other around" in a variety of fascinating patterns. Operation is as follows:

   - First the desired bit pattern is placed at the inputs of IC1 and IC2 with DIP switches S_1–S_8. Next, S_9, a normally closed push-button switch, is momentarily pressed. As a result, both the S_0 and S_1 mode controls are HIGH (S_0 being tied HIGH) and a parallel load takes place. The output LEDs reflect the input bit pattern.
   - The 555 clock generator now sends a 5-Hz signal to the clock inputs of both shift registers. With every clock pulse, the output bit pattern is shifted one position to the right. Since the Q_3 output of IC1 is fed to the DSR input of IC2, and the Q_3 output of IC2 is sent back to the DSR input of IC1, the bit pattern is recirculated.

   a. Referring to Figure 17.2, record missing IC pin numbers. Construct the circuit of Figure 17.2. Apply power.

   b. Is a random pattern of lit LEDs moving to the right at a 5-Hz rate?

   _____

   c. Load a desired bit pattern via DIP switches S_1–S_8. Press and release S_9.

   Describe what happens: _____

   _____

   _____

   _____

   d. Repeat the above procedure a few times, programming different input bit patterns using the DIP switches.

SUMMARY

In this experiment you investigated operation of the TTL 74194 4-bit bidirectional universal shift register. First you saw how it operates in the serial mode to shift data to the right and to the left. Next you operated the 74194 in the parallel (broadside) load mode. You also saw how the inhibit clock mode functions. Finally, you built and then analyzed operation of a Programmable Sequence Generator using two cascaded 74194 ICs.

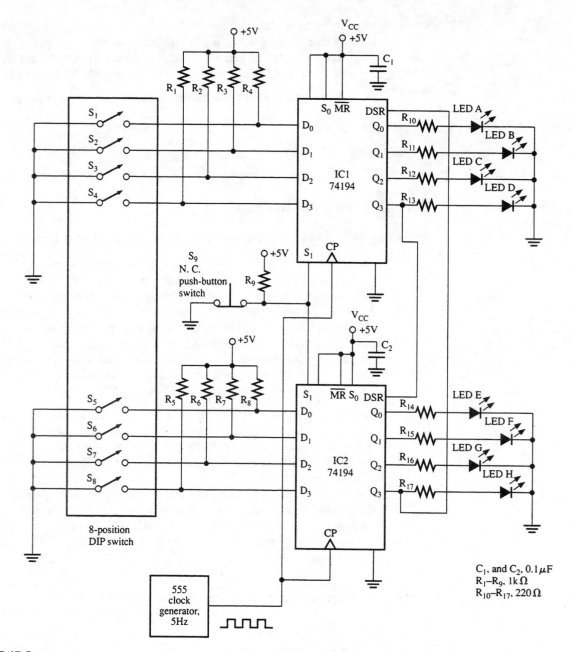

FIGURE 17.2

REVIEW QUESTIONS

1. Why is the TTL 74194 known as a universal shift register? _____
   _____
   _____

2. With the 74194, clocking takes place on the transition from a _____ to a
   _____ level.

3. What mode is the 74194 in when both S_0 and S_1 are HIGH? _____

4. Draw the block diagram for a 16-LED programmable sequence generator circuit similar to the one in Figure 17.2.

Writing Skills Assignment

In as few paragraphs as possible, explain how the Programmable Sequence Generator Circuit works. You might first discuss what the circuit does and then discuss how it does it.

18 Asynchronous Counters

OBJECTIVES

After completing this experiment, you will be able to

- Verify operation of a TTL 7493 binary asynchronous counter
- Verify operation of a TTL 7490 decade asynchronous counter
- Cascade binary and decade counters
- Build and analyze a Photoelectric Counter Circuit that uses two 7490 decade counters

REFERENCE READING

Review Ronald Reis, *Digital Electronics Through Project Analysis*, Chapter 11, Section 11.2.

EQUIPMENT & MATERIALS NEEDED

Equipment

☐ 1 5-V power supply
☐ 1 logic pulser
☐ 1 solderless circuit board

Materials

☐ 1 7400 quad two-input NAND gate
☐ 2 7490 decade counters
☐ 1 7493 binary counter
☐ 1 2N3904 NPN transistor
☐ 8 LEDs (red)
☐ 8 220-Ω resistors
☐ 2 1-kΩ resistors
☐ 1 100-kΩ resistor
☐ 1 100-kΩ potentiometer
☐ 1 photoresistor
☐ 2 N.O. push-button switches
☐ 1 package of jumper wires

BACKGROUND INFORMATION

Asynchronous means "not in sync," that is, "not in time." An asynchronous counter is one in which the clock inputs of each stage are not synchronized. Instead, the Q (or $\overline{Q}$) output of the first counter stage (flip-flop) is fed to the clock input of the second stage; the Q output of the second stage is fed to the clock input of the third stage, and so on. Hence the count "ripples" from one stage to the next. The process is illustrated in Figure 18.1a.

Asynchronous counters are fine as long as the circuitry is simple and relatively low-speed operation is acceptable. The main problem with such counters is that it takes time for a flip-flop to set or reset and produce a clock signal for the following flip-flop. If enough flip-flops are used, the propagation delay from the first to the last flip-flop can be substantial, often creating an intolerable error.

A synchronous counter, on the other hand, has the clock inputs of all its stages tied together. Hence there is no ripple effect; all flip-flop outputs react at the same time. The result is a counter where maximum speed is not held in check by propagation delay. The synchronous counter is illustrated in Figure 18.1b.

In this experiment you will examine asynchronous counters. You will investigate the ubiquitous TTL 7493 binary and TTL 7490 decade counters. You will see how such counters are cascaded and then conclude by investigating a Photoelectric Counter Circuit utilizing the cascading principle.

PROCEDURE

Part 1: Circuit Fundamentals

Binary and Decade Counters

☐ 1. You have already experimented with an asynchronous binary counter in Experiment 16 (refer to Figure 16.5). At that time you cascaded two 7473 dual J-K flip-flops to form a divide-by-16 binary counter. Such a counter, however, already comes packaged in a single IC, the TTL 7493, the pin configuration of which is shown in Appendix A (Figure A.16).

Actually, the 7493 is not only a divide-by-16 counter, but a divide-by-2 and a divide-by-8 counter as well. In its basic form, the chip contains four J-K flip-flops connected as shown in Figure 18.2. If the pulses to be counted are sent to clock input A ($\overline{CP}_0$), and the output taken at Q_0, the output frequency is

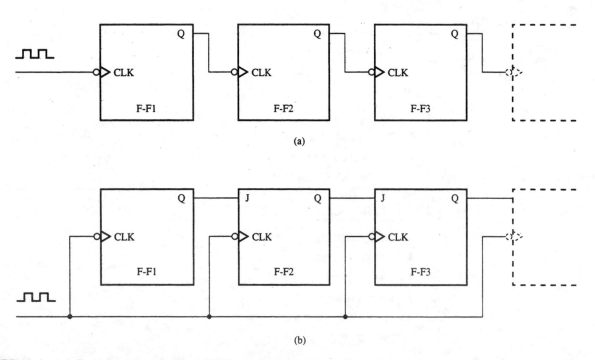

(a)

(b)

FIGURE 18.1 (a) Asynchronous counter and (b) synchronous counter.

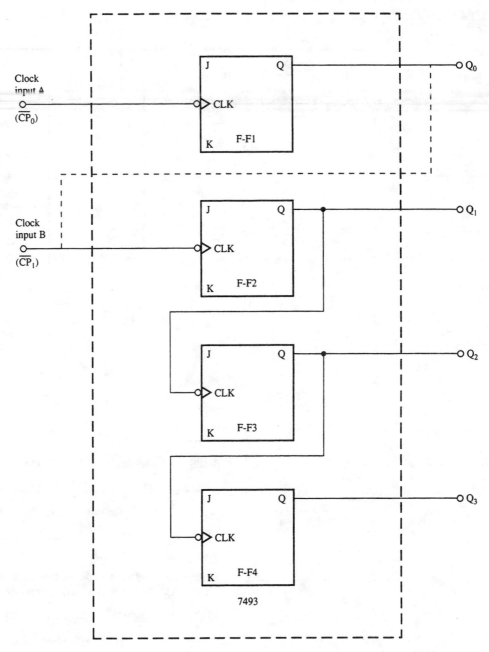

FIGURE 18.2

one-half that of the input frequency. Thus you have a divide-by-2 counter. On the other hand, if pulses are placed on clock input B ($\overline{CP}_1$), and the output taken from Q_3, the output frequency is one-eighth that of the input frequency since flip-flops 2–4 are cascaded. Finally, if an input signal is placed on clock input A, and the output at Q_0 is fed to clock input B (dotted lines), all four flip-flops are cascaded. If the output is taken from Q_3, the output frequency is one-sixteenth that of the input frequency. You have a divide-by-16 counter.

☐   a. Referring to Figure 18.3, record missing IC pin numbers. Construct the 7493 test circuit of Figure 18.3. Do not connect a jumper wire from Q_0 to $\overline{CP}_1$ (dotted lines) yet. Apply power.

☐   b. Using your logic pulser, "pulse" clock input A ($\overline{CP}_0$) a few times. Does

LED D_1 light? _____ Do LEDs D_2–D_4 light? _____ Is

the circuit dividing by 2? _____ How do you know? _____

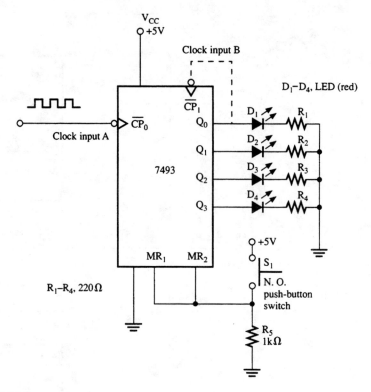

FIGURE 18.3

□ c. Using your logic pulser, "pulse" clock input B ($\overline{CP}_1$) a few times. Does
   LED D_1 light? _____ Do LEDs D_2–D_4 light in binary sequence?
   _____ Is the circuit dividing by 8? _____ How do you
   know? _____

   Is the circuit performing as an octal counter? _____

□ d. Connect a jumper wire from pin Q_0 to $\overline{CP}_1$. Using your logic pulser,
   "pulse" clock input A ($\overline{CP}_0$) at least 16 times. Do LEDs D_1–D_4 light
   in binary sequence? _____ Is the circuit dividing by 16?
   _____ How do you know? _____

   Is the circuit performing as a hexadecimal counter? _____

□ e. Using your logic pulser, advance the count to any binary number. Press
   S_1. Does the count return to zero? _____ The MR_1 and MR_2
   inputs are reset-to-zero inputs. When both are brought _____,
   the 7493 resets to zero.

□ 2. The 7490, the pin configuration of which is shown in Appendix A (Figure A.15),
   is similar to the 7493, only the former is a decade counter. Actually, the 7490 is
   also a divide-by-2 and a divide-by-5 counter. Its internal flip-flop structure is
   comparable to that of the 7493, though additional gating is required to create a
   decade counter.

   □ a. Referring to Figure 18.4, record missing IC pin numbers. Construct the
      7490 test circuit of Figure 18.4. Do not connect a jumper wire from Q_0
      to $\overline{CP}_1$ (dotted line) yet. Apply power.

   □ b. Using your logic pulser, "pulse" clock input A ($\overline{CP}_0$) a few times. Does
      LED D_1 light? _____ Do LEDs D_2–D_4 light? _____

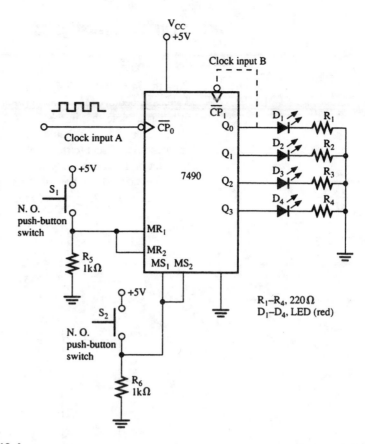

FIGURE 18.4

Is the circuit dividing by 2? _____ How do you know?

_____

☐ c. Using your logic pulser, "pulse" clock input B ($\overline{CP}_1$) a few times. Does LED D_1 light? _____ Do LEDs D_2–D_4 light in binary sequence? _____ Is the circuit dividing by 5? _____ How do you know? _____

Is the circuit performing as a penta (5) counter? _____

☐ d. Connect a jumper wire from Q_0 to $\overline{CP}_1$. Using your logic pulser, "pulse" clock input A ($\overline{CP}_0$) at least 10 times. Do LEDs D_1–D_4 light in binary sequence? _____ Is the circuit dividing by 10? _____ How do you know? _____

Is the circuit performing as a decade counter? _____

☐ e. Using your logic pulser, advance the count to any binary number. Press S_1. Does the count return to zero? _____ The MR_1 and MR_2 inputs are reset-to-zero inputs. When brought _____, the 7490 resets to zero.

☐ f. Using your logic pulser, advance the count to any binary number other than 9. Press S_2. Does the count go to 9? The MS_1 and MS_2 inputs are reset-to-9 inputs. When brought _____, the 7490 resets to 9.

Part 2: Further Investigation

Cascading Counters

☐ 1. Both the 7493 and the 7490 counters can be cascaded to increase the total count. Two cascaded decade counters (Figure 18.5) will reach a count of 99 (1001 1001 in binary). Note that in the figure the Q_3 output of IC1 is connected to the $\overline{CP}_0$ clock input of IC2. When a tenth clock pulse is received at IC1's clock input, another pulse is sent to the clock input of IC2, causing its Q_0 output to go HIGH. A count of 10 is now indicated.

To understand how this expanded counting circuit works, refer to the simplified timing diagram of Figure 18.6. Keep in mind that a count advances

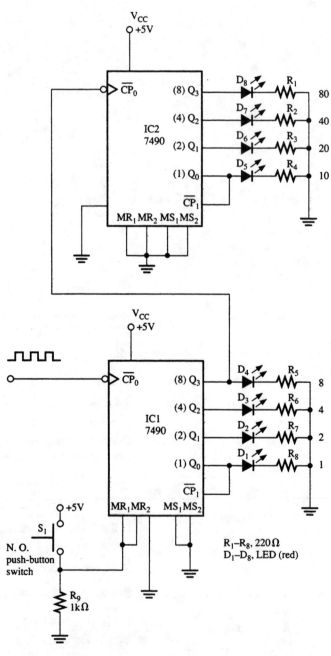

FIGURE 18.5

156

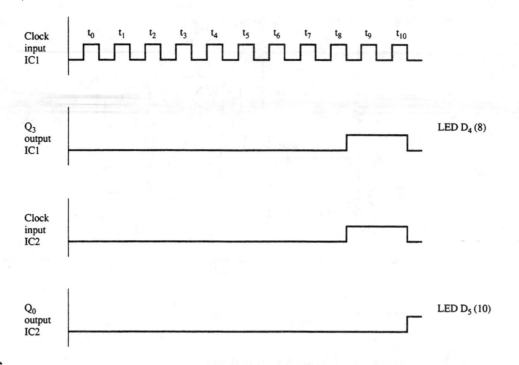

FIGURE 18.6

only on the application of a *negative*-going clock pulse. Understanding this is the key to comprehending the cascading process. Here is how the circuit works:

- When the eighth clock pulse to IC1 transitions from a HIGH to a LOW, the Q_3 output of IC1 goes HIGH. LED D_4 turns on. Also, the clock input to IC2 goes HIGH; however, this has no effect on the Q_0 output of IC2 because the clock input to IC2 has not transitioned from a HIGH to a LOW yet. Thus the Q_0 output of IC2 remains LOW (LED 5 is off).
- When the tenth clock pulse to IC1 transitions from a HIGH to a LOW, the Q_3 output of IC1 goes LOW, and LED D_4 turns off. The clock input to IC2 also goes LOW. When that happens, the Q_0 output of IC2 goes HIGH. LED D_5 turns on, indicating a 10.

☐ a. Referring to Figure 18.5, record missing IC pin numbers. Construct the circuit of Figure 18.5. Apply power.

☐ b. Press S_1 to reset both ICs to zero. Are any LEDs on? _____

☐ c. Using your logic pulser, "pulse" the clock input to IC1 until the count reaches 8. Which LEDs are on? _____ Pulse the clock input again. Which LEDs are on? _____ Pulse the clock input a third time. Which LED(s) are on? _____ What is the count? _____

☐ d. Continue to pulse the clock input to IC1 until the circuit's maximum count is reached. Which LEDs are on? _____. What is the count? _____ Pulse the clock input again. What LEDs are on? _____ Why is this so? _____
_____

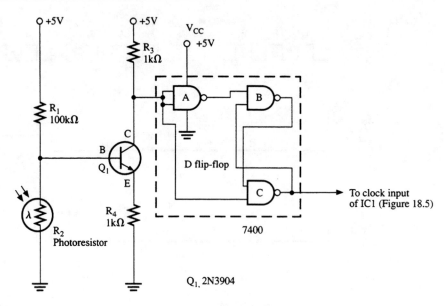

FIGURE 18.7

Part 3: Circuit Applications

A Two-Stage Counter Circuit: Photoelectric Counter Circuit

☐ 1. By adding the photoelectric trigger circuit of Figure 18.7 to the two-stage decade counter circuit of Figure 18.5, you can create a Photoelectric Counter. Such a circuit will count the number of times a light beam directed on the photoresistor is interrupted.

With light shining on the photoresistor, its resistance is LOW (around $500\,\Omega$), the base of Q_1 is near ground, the transistor is at cutoff, and the input to the D flip-flop is HIGH. When light to the photoresistor is interrupted, the photoresistor resistance shoots up (to over $500\,k\Omega$), the base of Q_1 is near $+5\,V$, the transistor conducts, and the input to the D flip-flop goes LOW. Resistor R_4 acts as a sensitivity control. Thus with each interruption of the light beam, a negative pulse is created at the input of the D flip-flop. A similar pulse appears at the output of the flip-flop and is sent to the clock input of IC1.

☐ a. Referring to Figure 18.7, record missing IC pin numbers. Construct the circuit of Figure 18.7 and connect its output to the clock input of IC1 in the circuit of Figure 18.5. Apply power.

☐ b. Shine a light directly on the photocell from about a foot away. Bring a sheet of cardboard (or similar material) across the photocell, interrupting the light beam. As you do, note if the circuit counts. You might have to play with the distance between the cardboard sheet and the photocell and adjust R_4 to get the circuit to count. Once counting takes place, reset the circuit by pressing S_1 (Figure 18.5). Now, using the cardboard to break the light beam, advance the count to its maximum of 99 ($1001\ 1001_2$).

SUMMARY

In this experiment you investigated asynchronous counters. You studied operation of the TTL 7493 binary counter and the TTL 7490 decade counter. You cascaded two 7490s and concluded by building, then analyzing operation of, a Photoelectric Counter Circuit.

REVIEW QUESTIONS

1. Explain the difference between an asynchronous and a synchronous counter. _____

_____

_____

2. The 7490 and the 7493 ICs have _____ edge-triggered clocks.
3. Draw the schematic for a circuit that will count to 999 in binary and use three 7490 decade counters.

4. Draw a schematic showing how the 7493 can be used as a divide-by-10 counter.

Writing Skills Assignment

In as few paragraphs as possible, explain how the Photoelectric Counter Circuit works. You might first discuss what the circuit does and then discuss how it does it.

19 Synchronous Counters

OBJECTIVES

After completing this experiment, you will be able to

- Verify operation of a TTL 74193 up/down binary synchronous counter
- Assemble a 74193-based circuit to count up or down from N and recycle
- Assemble a 74193-based circuit to count up to N and recycle
- Build and analyze operation of an LED "Night Rider" Circuit using the 74193 as an up/down counter

REFERENCE READING

Review Ronald Reis, *Digital Electronics Through Project Analysis,* Chapter 11, Section 11.3.

EQUIPMENT & MATERIALS NEEDED

Equipment

- ☐ 1 5-V power supply
- ☐ 1 logic pulser
- ☐ 1 solderless circuit board

Materials

- ☐ 1 555 timer
- ☐ 1 7400 quad two-input NAND gate
- ☐ 1 7404 hex inverter
- ☐ 1 74154 4-line-to-16 line decoder/demultiplexer
- ☐ 1 74193 synchronous up/down 4-bit binary counter
- ☐ 16 LEDs (red)
- ☐ 1 0.47-µF capacitor
- ☐ 6 220-Ω resistors
- ☐ 6 1-kΩ resistors
- ☐ 1 100-kΩ potentiometer
- ☐ 2 N.O. push-button switches
- ☐ 1 8-position DIP switch (or equivalent)
- ☐ 1 package of jumper wires

BACKGROUND INFORMATION

Synchronous means "in sync," that is, "in time." Synchronous counters have the clock inputs to all their internal stages tied together; thus all stages operate in sync. Such counters are extremely fast and glitch-free. (For a simple comparison of asynchronous and synchronous counter operation, see Figure 18.1 in Experiment 18.)

Most counters are synchronous. Some count up, some count down, and a number count up and down. Of the latter, the TTL 74193 synchronous up/down 4-bit binary counter with dual clock, the pin configuration of which is shown in Appendix A (Figure A.25), is one of the most widely used. It will count up or down from any binary number 0 through 15, it has a fully independent clear input, and it contains internally provided cascading circuitry.

In this experiment you will use the 74193 to examine basic synchronous counter operation. You will investigate all the chip features and discover how the IC can be made to count up or down from any binary number, 0000–1111, and recycle. You will conclude by building and then analyzing operation of a most interesting 74193 application, an LED "Night Rider" Circuit.

PROCEDURE

Part 1: Circuit Fundamentals

Up/Down Counter Basics

☐ 1. Referring to Figure 19.1, we see that the 74193 IC has four binary-weighted outputs: Q_0, Q_1, Q_2, and Q_3. Each output is triggered by a LOW-to-HIGH transition of either clock input. If the count-down clock input is held HIGH and

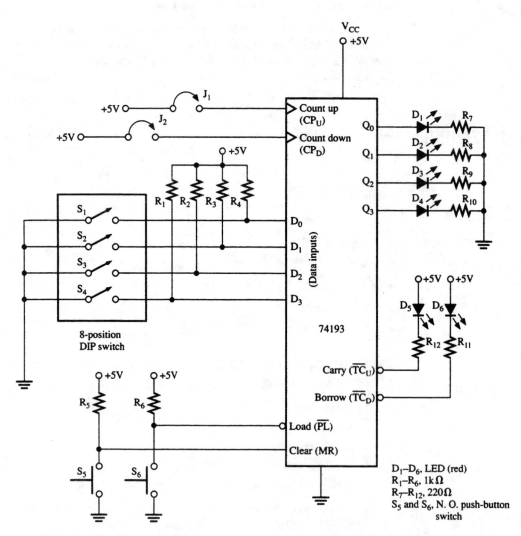

FIGURE 19.1

162

pulses are applied to the count-up clock input, the circuit counts up. Conversely, if the clock-up clock input is held HIGH while pulses are fed to the count-down clock input, the circuit counts down.

The 74193 is also fully programmable; that is, each output may be preset to a LOW or a HIGH level by entering the desired data at inputs D_0, D_1, D_2, and D_3 while bringing the load input ($\overline{PL}$) LOW. Thus the count length can be modified with the preset inputs.

A clear input, or master reset (MR), is also provided. When it is brought HIGH, all outputs are forced LOW.

Furthermore, the 74193 is designed for easy cascading with the use of carry and borrow outputs. The carry output ($\overline{TC_U}$) produces a pulse equal in width to the count-up input when an *overflow* condition exists. Similarly, the borrow output ($\overline{TC_D}$) produces a pulse equal in width to the count-down input when the counter *underflows*. Thus to cascade *up counters*, the carry output of one counter is fed to the count-up input of a succeeding counter. To cascade *down counters*, the borrow output of one counter is fed to the count-down input of a succeeding counter.

☐ a. Referring to Figure 19.1, record missing IC pin numbers. Construct the 74193 test circuit of Figure 19.1. Do not apply power yet.

☐ b. Open switches S_1–S_4 and jumpers J_1 and J_2. Apply power.

☐ c. Are any LEDs, D_1–D_4, on? _____ If so, which ones? _____ Close J_2, bringing the count-down input (CP_D) HIGH. Using your logic pulser, "pulse" the count-up input (CP_U) a few times. Is the circuit counting up? _____ What is the status of the "carry" ($\overline{TC_U}$) LED, D_5? _____ Continue advancing the count until it reaches 1110. Pulse the circuit one more time. What count is indicated? _____ What is the status of LED D_5? _____ Pulse the circuit one more time. What is the status of LEDs D_1–D_4? _____ What is the status of LED D_5? _____ Repeat the above procedure a few times. At any given count, press S_6, bringing the clear (MR) input momentarily HIGH. What happens to the output count? _____ Summarize your observations:

_____

_____

_____

☐ d. Advance the count to 1111. Close J_1, bringing the count-up input ($\overline{CP_U}$) HIGH. Using your logic pulser, "pulse" the count-down ($\overline{CP_D}$) input a few times. Is the circuit counting down? _____ What is the status of the "borrow" ($\overline{TC_D}$) LED, D_6? _____ Continue pulsing the circuit until the count reaches 0001. Pulse the circuit one more time. What count is indicated? _____ What is the status of LED D_6? _____ Pulse the circuit one more time. What is the status of LEDs D_1–D_4? _____ What is the status of LED D_6? _____ Repeat the above procedure a few times. Summarize your observations:

_____

_____

_____

☐ 2. The 74193 can be made to count up or down from a preset number.

☐ a. Close J_2, bringing the count-down ($\overline{CP_D}$) input HIGH. Set switches S_1–S_4 to produce 0111 (an arbitrarily chosen number) on the data inputs. Press and release S_5, bringing the load ($\overline{PL}$) input momentarily LOW.

What is the output count? _____ Using your logic pulser, "pulse" the count-up input $\overline{CP_U}$) until the count reaches 1111. Pulse the circuit one more time. What count is indicated? _____ Why is this so? _____

_____

☐ b. Press and hold S_5 closed. Pulse the circuit a dozen or so times. Is the count sequence different from that in step 2a? _____ In what way? _____

_____

Why is the count sequence different? _____

_____

☐ c. Place J_1 HIGH. With S_1–S_4 still set to produce 0111 on the data inputs, press and release S_5. What is the count output? _____ Using your logic pulser, "pulse" the count-down input until the count reaches 0000. Pulse the circuit one more time. What count is indicated? _____ Why is this so? _____

_____

☐ d. Press and hold S_5 closed. Pulse the circuit a dozen or so times. Is the count sequence different from that in step 2c? _____ In what way? _____

_____

Why is the count sequence different? _____

_____

Part 2: Further Investigation

Counting Up or Down From Any Desired N

☐ 1. The 74193 up/down counter can be made to count up or down from any desired N (number) and recycle. To test this, we modify the circuit of Figure 19.1 as shown in Figure 19.2. (Note that S_6, R_6, R_{11}, R_{12}, LED D_5, and LED D_6 are removed.)

FIGURE 19.2

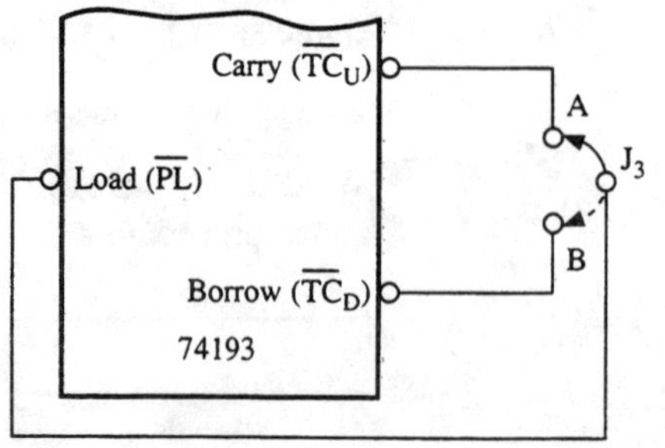

164

To have the circuit count up from any N and recycle, J_3 is placed in position A. When the count reaches 1111 and then overflows, the negative carry pulse loads N (by bringing the clear, MR, input of the 74193 momentarily HIGH) and the count recycles.

To cause the circuit to count down from any N and recycle, J_3 is placed in position B. When the count reaches 0000 and then underflows, the negative borrow pulse ($\overline{TC}_D$) loads N and the count recycles.

☐   a. Modify the circuit of Figure 19.1 as shown in Figure 19.2. Apply power.

☐   b. Close J_2. (J_1 should be open.) Place J_3 in position A. Enter any binary number using switches S_1–S_4. Using your logic pulser, "pulse" the count-up input enough times to determine the counting sequence. Is the circuit counting up from the number you entered and recycling?

_____ Does the count ever reach 1111? _____ Why is

this so? _____

_____

☐   c. Close J_1. (J_2 should be open.) Place J_3 in position B. Enter any binary number using switches S_1–S_4. Using your logic pulser, "pulse" the count-down input enough times to determine the counting sequence. Is the circuit counting down from the number you entered and recycling?

_____ Does the count ever reach 0000? _____ Why is

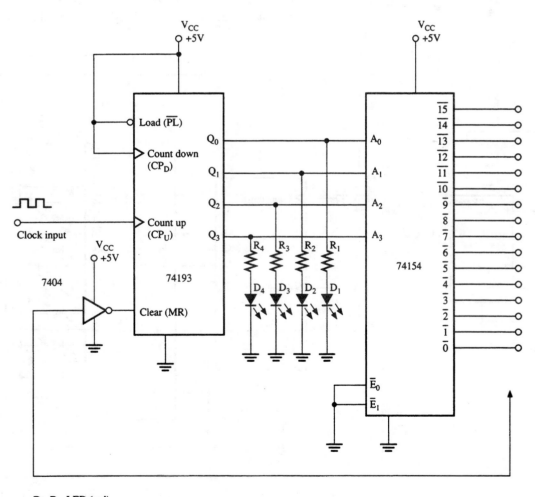

D₁–D₄, LED (red)
R₁–R₄, 220 Ω

FIGURE 19.3

this so? _____

☐ 2. The 74193 can also be made to count up *to* N and recycle. A circuit that will do this is shown in Figure 19.3. As the count-up input on the 74193 input is pulsed, one output at a time on the 74154 goes LOW in sequence. That LOW is converted to a HIGH by the 7404 NOT gate—the 74193 is then cleared to start its count at zero again. LEDs D_1–D_4 indicate the binary count.

  ☐ a. Referring to Figure 19.3, record missing IC pin numbers. Construct the circuit of Figure 19.3. Apply power.

  ☐ b. Connect the input of the NOT gate to output $\overline{4}$ of the 74154. Using your logic pulser, ''pulse'' the count-up input a few times while observing

  LEDs D_1–D_4. Is the circuit operating as expected? _____

  ☐ c. Repeat the above procedure, placing the input of the NOT gate at selected outputs of the 74154. Pulse the count-up input enough times to determine

  if the circuit is functioning as it should. Is it? _____

Part 3: Circuit Applications

Counting Up and Down: LED ''Night Rider'' Circuit

☐ 1. An application that uses the 74193 to count both up and down is shown in Figure 19.4. This LED ''Night Rider'' Circuit (from the television series of the same name) causes a row of 16 LEDs to ''swing'' back and forth with only one LED

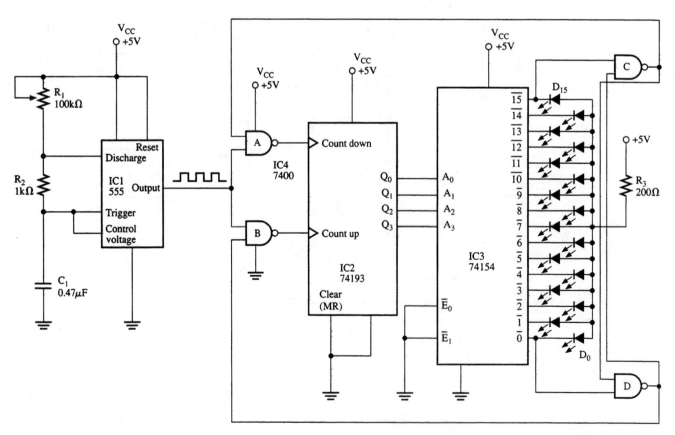

D_0–D_{15}, LED (red)

FIGURE 19.4

166

on at a time. As we can see from the schematic, the circuit consists of a 555 IC operating as an astable multivibrator, a 74193 binary up/down counter, a 74154 1-of-16 data distributor, and four NAND gates in a 7400 IC. Here is how the circuit works:

- Recall that all outputs of the 74154 are HIGH except one.
- Assume that output $\overline{0}$ of the 74154 is LOW. Thus its accompanying LED (D_0) is lit. In addition, the lower input pin of NAND gate D is LOW and its output is HIGH. With the output of NAND gate D HIGH, both inputs to NAND gate C are HIGH and its output is LOW. The upper input pin of NAND gate D is also LOW, confirming the gate's HIGH output.
- With the outputs of gates C and D LOW and HIGH, respectively, the upper input pin of gate A is LOW and the lower input pin of gate B is HIGH.
- Clock pulses arriving from the 555 are gated through NAND gate B to the count-up input of IC2. With a LOW on the upper input pin of gate A, however, its output is held HIGH, as is the count-down input of IC2. The 74193 now begins to count up with each clock pulse.
- Even though output $\overline{0}$ of IC2 is no longer LOW as the count progresses, the outputs to both NAND gates C and D remain as before. Thus the 74193 is kept counting up. The LEDs "advance," lighting in sequence.
- When output $\overline{15}$ of IC3 goes LOW, LED D_{15} turns on. Consequently, with the upper input pin of NAND gate C LOW, its output goes HIGH, while the output of NAND gate D goes LOW. Now clock pulses are inhibited from getting through gate B to the count-up input of IC2. But they do get through gate A (its upper input pin is now HIGH) to the count-down input of the 74193. The LEDs "reverse direction" as IC2 begins counting down.
- When LED D_0 lights again, the whole cycle is repeated.

  If you adjust potentiometer R_1, the LEDs' back-and-forth speed can be varied.

☐ a. Referring to Figure 19.4, record missing IC pin numbers. Construct the circuit of Figure 19.4. Apply power.

☐ b. Adjust R_1 until the LEDs "move" back and forth at an eye-pleasing rate.

Does every LED light in sequence? _____

SUMMARY

In this experiment you examined synchronous counter operation. You investigated the TTL 74193 synchronous up/down counter, exploring its many features. You concluded by building and then analyzing operation of an LED "Night Rider" Circuit.

REVIEW QUESTIONS

1. Both clocks of the 74193 are _____ edge-triggered.
2. Draw the circuit for two cascaded 74193 ICs that will count up or down.

3. For a 74193, regardless of the status of the data inputs and outputs, if the clear input (MR) goes _____, the data outputs all go _____.
4. For a 74193, data on the data inputs will appear on the data outputs only when the _____ input is _____.

Writing Skills Assignment

In as few paragraphs as possible, explain how the LED "Night Rider" Circuit works. You might first discuss what the circuit does and then discuss how it does it.

20 Digital-to-Analog (D/A) Conversion

OBJECTIVES

After completing this experiment, you will be able to

- Verify operation of a binary-weighted digital-to-analog (D/A) converter
- Verify operation of an R/2R ladder digital-to-analog converter
- Build and analyze operation of a DAC0808 8-Bit DAC Demonstration Circuit

REFERENCE READING

Review Ronald Reis, *Digital Electronics Through Project Analysis,* Chapter 12, Sections 12.1 and 12.2.

EQUIPMENT & MATERIALS NEEDED

Equipment

- ☐ 1 power supply with ground, +5, −12, and +12 V
- ☐ 1 dual-channel oscilloscope
- ☐ 1 voltmeter (preferably digital)
- ☐ 1 555 clock generator
- ☐ 1 solderless circuit board

Materials

- ☐ 1 741 operational amplifier
- ☐ 1 7447 BCD-to-seven-segment decoder/driver
- ☐ 1 7493 binary counter
- ☐ 1 DAC0808 8-bit D/A converter
- ☐ 1 common-anode seven-segment LED display
- ☐ 1 0.01-μF capacitor
- ☐ 7 220-Ω resistors
- ☐ 10 1-kΩ resistors
- ☐ 4 2-kΩ resistors
- ☐ 4 10-kΩ resistors
- ☐ 1 12-kΩ resistor
- ☐ 6 15-kΩ resistors

☐ 7 100-kΩ resistors

☐ 1 8-position DIP switch (or equivalent)

☐ 1 package of jumper wires

BACKGROUND INFORMATION

In the "real world," various phenomena take place in an analog manner. Temperature rises and falls slowly; light increases and decreases in intensity gradually; motion is not abrupt, but occurs in a continuous flow. The transducers that sense these phenomena on the input end of an electrical system and those that produce such phenomena on the output end create and require, respectively, analog signals. An input transducer that changes light, heat, sound, or mechanical motion to electricity will produce a gradually varying output signal that is proportional to the regularly changing input condition. The opposite is true when an output transducer is required to create varying amounts of light, heat, sound, or mechanical motion. Such a transducer must receive a varying electrical signal that is proportional to the output energy demanded.

If signals within a circuit are to be manipulated digitally, however, the analog output of the input transducer must be converted to a corresponding digital signal before it can be processed by the main circuit. That is the function of the *analog-to-digital converter (ADC)*. If the output transducer requires an analog signal, a *digital-to-analog converter (DAC)* is placed between the main circuit and the output transducer to convert digital signals to analog. While the addition of ADCs and DACs means added system complexity, digital signal processing, especially within microprocessors, offers such tremendous advantages that the additional circuitry is usually well worth the effort.

In this experiment, you will examine three DAC circuits. First you will explore the binary-weighted D/A converter that uses "weighted" resistor values in combination with an op amp to produce a simple and effective DAC. Next, you will move on to the more advanced R/2R ladder DAC, seeing first how the ladder itself operates, then how a counter can be incorporated to create the familiar analog staircase output. Finally, you will investigate a DAC0808 8-Bit DAC Demonstration Circuit, known as a "multiplying DAC." You will see how this industry-standard 16-pin device performs complex digital-to-analog conversion, all for a cost of less than $2.

PROCEDURE

Part 1: Circuit Fundamentals

Binary-Weighted D/A Converter

☐ 1. The binary-weighted D/A converter is the simplest of all DACs. Such a converter is shown in Figure 20.1. This 4-bit binary-weighted circuit will recognize a binary number, 0000–1111, and convert its value into a corresponding analog voltage.

Note that there are four standard binary inputs: A, B, C, and D. When switch S_1 is closed, current flows through it and R_1 to the virtual ground at the inverting input (−) of the 741 op amp. (See the pin diagram for the 741 op amp in Figure A.29, Appendix A.) The same is true for switches S_2–S_4 and their respective resistors, R_2–R_4. (Note the parallel and series/parallel resistor combinations required to arrive at the desired values for R_2, R_3, and R_4.) A closed switch represents a 1, an open switch, a 0 for a given input.

The inputs, A–D, are said to be "weighted" because their respective resistors are weighted. With weighted resistors, a weighted current will flow through them when their respective switches are closed. The result will be a weighted current through R_f and a corresponding weighted voltage drop at the op-amp output. Since the circuit also *sums* the currents arriving at the op amp (−), we have a D/A converter that will recognize 16 different binary numbers (current) at the input and produce a specific related voltage at the output. The corresponding voltages for each binary number, 0000–1111, are shown in Table 20.1.

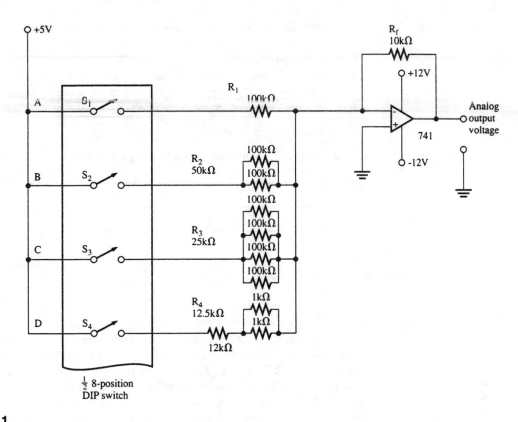

FIGURE 20.1

TABLE 20.1

S₄ D	S₃ C	S₂ B	S₁ A	V_out calculated	V_out measured
0	0	0	0	0	
0	0	0	1	−0.5	
0	0	1	0	−1	
0	0	1	1	−1.5	
0	1	0	0	−2	
0	1	0	1	−2.5	
0	1	1	0	−3	
0	1	1	1	−3.5	
1	0	0	0	−4	
1	0	0	1	−4.5	
1	0	1	0	−5	
1	0	1	1	−5.5	
1	1	0	0	−6	
1	1	0	1	−6.5	
1	1	1	0	−7	
1	1	1	1	−7.5	

☐ a. Referring to Figure 20.1, record missing IC pin numbers. Construct the circuit of Figure 20.1. Place a voltmeter at the output of the circuit. Apply power.

☐ b. Set up the input switch conditions shown in Table 20.1 and record the measured analog output voltage levels in the space provided.

☐ c. Are the measured analog output voltages consistent with the calculated analog output voltages? _____

☐ d. Summarize your conclusions with regard to Part 1:_____

_____

_____

_____

Part 2: Further Investigation

R/2R Ladder DAC

☐ 1. Given the inherent disadvantages of the binary-weighted D/A converter (odd resistor values, for example), most DAC circuits are instead composed of what is known as an R/2R ladder. Such a circuit consists of only two resistor values, R and 2R, arranged in the form of a ladder, as shown in Figure 20.2. In the R/2R ladder DAC, the analog output voltage will always be proportional to the binary number set on the input switches. The formula for determining output voltage is

$$\begin{array}{cccc} \text{MSB} & & & \text{LSB} \\ \text{(D)} & \text{(C)} & \text{(B)} & \text{(A)} \end{array}$$

$$V_{out} = \frac{V_{ref}}{2} = \frac{V_{ref}}{4} = \frac{V_{ref}}{8} = \frac{V_{ref}}{16}$$

where V_{out} = analog output voltage
V_{ref} = reference voltage

For example, if the reference voltage is +5 V, and switches S_2 and S_3 are set HIGH while switches S_1 and S_4 are set LOW (binary 0110), then

$$\begin{array}{cc} \text{(A)} & \text{(B)} \end{array}$$

$$\begin{aligned} V_{out} &= \frac{5}{4} + \frac{5}{8} \\ &= 1.25 + 0.625 \\ &= 1.875 \text{ V} \end{aligned}$$

(*Note:* Only the terms for which there is a HIGH switch setting—in this example, inputs B and C—are included.)

☐ a. Construct the circuit of Figure 20.2. Place a voltmeter at the output of the circuit. Apply power.

☐ b. Calculate the analog output voltages for the binary numbers 0000–1111 and record the results in the space provided in Table 20.2.

☐ c. Set up the input switch conditions shown in Table 20.2 and record the measured analog output voltage levels in the space provided.

☐ d. Are the measured analog output voltages consistent with the calculated analog output voltages? _____

☐ e. Summarize your conclusions with regard to this first half of Part 2: ___

_____

_____

_____

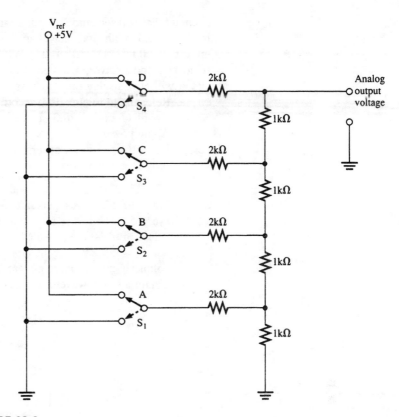

FIGURE 20.2

TABLE 20.2

2 D	4 C	8 B	16 A	V_out calculated	V_ref measured
0	0	0	0		
0	0	0	1		
0	0	1	0		
0	0	1	1		
0	1	0	0		
0	1	0	1		
0	1	1	0		
0	1	1	1		
1	0	0	0		
1	0	0	1		
1	0	1	0		
1	0	1	1		
1	1	0	0		
1	1	0	1		
1	1	1	0		
1	1	1	1		

2. A circuit that will sequentially address inputs to the 4-bit binary ladder is shown in Figure 20.3. With clock pulses arriving at IC1, the 7493 binary counter counts up from 0000 to 1111 and repeats. With the use of IC2, a 7447 BCD-to-seven-segment decoder/driver, and a common-anode seven-segment LED readout, the circuit will now display the "address" selected by IC1.

   a. Referring to Figure 20.3, record missing IC pin numbers. Construct the circuit of Figure 20.3. Place a voltmeter at the output of the circuit. Apply power.

   b. Using the 555 clock generator, place a 1-Hz signal at the clock input to IC1. Does the display indicate the symbols 0, 1, 2, 3, 4, 5, 6, 7, 8, 9,

   $\sqsubset$, $\sqsupset$, $\sqcup$, $\underline{\smile}$, $\underline{\angle}$, and a blank in succession? _____

   c. Slow the 555 clock generator to produce a pulse approximately once every 5 s. Note the measured analog output voltages and record the values in the space provided in Table 20.3.

   d. Set the 555 clock generator to approximately 1 kHz. Remove the voltmeter and connect an oscilloscope to the analog voltage output. Obtain a stable waveform on the oscilloscope screen.

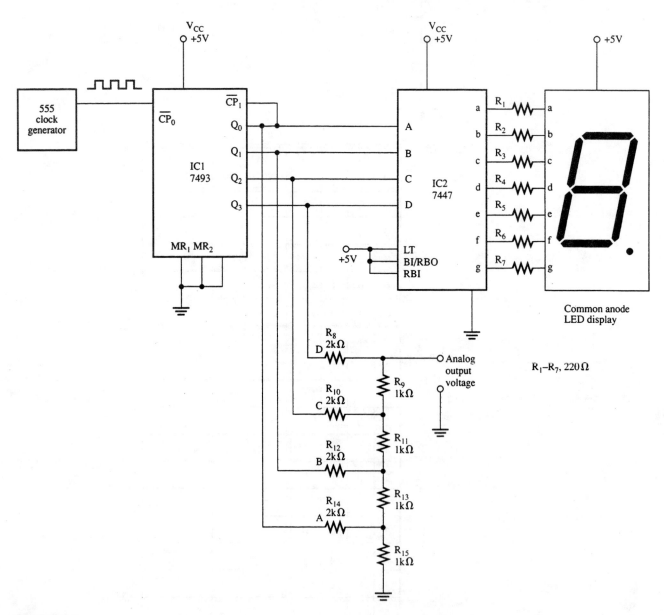

FIGURE 20.3

174

TABLE 20.3

Address	Binary number	Analog output voltage
0	0000	
1	0001	
2	0010	
3	0011	
4	0100	
5	0101	
6	0110	
7	0111	
8	1000	
9	1001	
⊏	1010	
⊐	1011	
⊔	1100	
⩵	1101	
⩵	1110	
Blank	1111	

☐ e. Draw the waveform displayed on the oscilloscope in Plot 20.1. Summarize your conclusions with regard to the second half of Part 2:

_____

_____

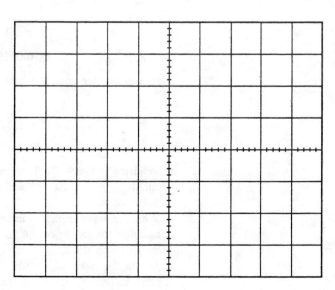

PLOT 20.1

Part 3: Circuit Applications

A DAC IC: DAC0808 8-Bit DAC Demonstration Circuit

☐ 1. You do not have to build a D/A converter with discrete resistors and op amps. A number of DAC ICs are available with a host of useful features, many selling for less than $2. One such chip, the 8-bit DAC0808, is obtainable from several sources, including Precision Monolithics, National Semiconductor, and Motorola. The pin diagram for the DAC0808 is shown in Appendix A (Figure A.31).

A DAC0808 8-Bit DAC Demonstration Circuit, using the DAC0808 IC, is shown in Figure 20.4. Two types of signals are required to make the DAC0808 function: an analog reference signal and a binary number (digital word). The device will produce an output current that is the product of these two factors; hence it is often called a *multiplying DAC*. It is important to note that the DAC0808 produces an output *current,* not an output voltage. To convert the current output to a voltage output, an op amp, acting as a current-to-voltage converter, is added at the DAC0808 output.

The current output of the DAC0808 (I_{out}) is determined by the formula

$$I_{out} = I_{ref} \times \frac{A}{2^n}$$

where I_{out} = output current of the DAC0808
  I_{ref} = DAC reference current (2 mA, or 12 V/6000, in the circuit of Figure 20.4)
  A = binary number at the DAC inputs in decimal form, 0–255
  2^n = bit resolution (256 for the 8-bit DAC0808)

For finding the analog output voltage (of the op amp) the formula is

$$V_{out} = I_{out} \times R_f$$

where V_{out} = analog output voltage of the op amp
  I_{out} = output current of the DAC0808
  R_f = feedback resistance

Let's take an example. For the circuit of Figure 20.4, what would the output voltage be if the binary number at the input is 10000101 (decimal 133)?

First you find the output current of the DAC0808:

$$I_{out} = I_{ref} \times \frac{A}{2^n}$$

$$= 2\,mA \times \frac{133}{256}$$

$$= 0.00103\,A\ (1.03\,mA)$$

Next you find the analog output voltage:

$$V_{out} = I_{out} \times R_f$$
$$= 1.03\,mA \times 2.5\,k\Omega$$
$$= 2.59\,V$$

☐ a. Referring to Figure 20.4, record missing IC pin numbers. Construct the circuit of Figure 20.4. Place a voltmeter at the output of the circuit. Apply power.

☐ b. With the input binary numbers shown in steps 1–4 of Table 20.4, calculate the analog output voltage for each step and record the value in the space provided.

☐ c. Set up the input conditions in steps 1–4 of Table 20.4 and, as you perform each step, record the measured output analog voltage in the space provided.

176

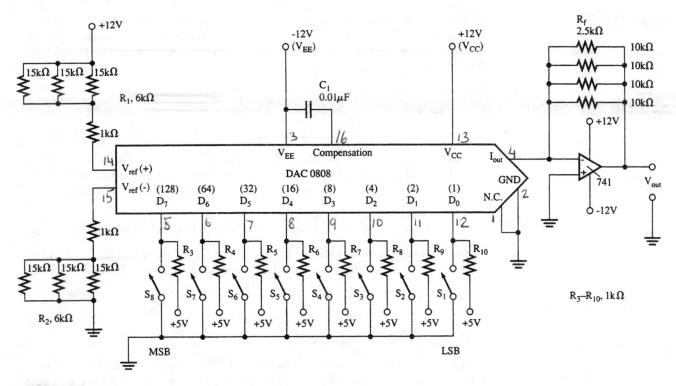

FIGURE 20.4

d. Select your own input binary numbers for steps 5 and 6 of Table 20.4. Calculate and record the analog output voltage expected. Set up the input conditions in steps 5 and 6 and, as you perform each step, record the measured analog output voltage in the space provided.

e. Are the measured analog output voltages consistent with the calculated analog output voltages? _____

f. Summarize your conclusions with regard to Part 3: _____

_____

_____

_____

Step	Input binary number	(decimal number)	Calculated output voltage	Measured output voltage
1	10010000	(44)		
2	01100101	(101)		
3	11001000	(200)		
4	01000100	(68)		
5				
6				

TABLE 20.4

In this experiment you examined digital-to-analog conversion. You began by investigating the binary-weighted D/A converter. Next you explored the R/2R ladder digital-to-analog converter. Finally, you built and then analyzed the operation of a DAC0808 8-Bit DAC Demonstration Circuit.

REVIEW QUESTIONS

1. Explain why the circuit of Figure 20.1 is known as a summing amplifier. _____

   _____

   _____

2. If the circuit of Figure 20.2 has a reference voltage of $+12$ V, and switches S_1 and S_2 are placed LOW while switches S_3 and S_4 are placed HIGH, what will the analog output voltage be? _____

3. In the circuit of Figure 20.4, what is the analog output voltage if switches S_1, S_3, S_5, and S_7 are closed while switches S_2, S_4, S_6, and S_8 are open? _____

4. In the circuit of Figure 20.4, what is the analog output voltage if all switches, S_1–S_8, are left open? _____

Writing Skills Assignment

In as few paragraphs as possible, explain how the DAC0808 8-Bit DAC Demonstration Circuit works. You might first discuss what the circuit does and then discuss how it does it.

21 Analog-to-Digital (A/D) Conversion

OBJECTIVES

After completing this experiment, you will be able to

- Verify operation of a simple flash analog-to-digital converter
- Verify operation of a flash analog-to-digital converter with BCD decoding
- Build and analyze operation of an ADC0804 8-Bit ADC Demonstration Circuit

REFERENCE READING

Review Ronald Reis, *Digital Electronics Through Project Analysis*, Chapter 12, Section 12.3.

EQUIPMENT & MATERIALS NEEDED

Equipment

☐ 1 5-V power supply
☐ 1 voltmeter (preferably digital)
☐ 1 solderless circuit board

Materials

☐ 3 LM339 quad comparators
☐ 1 74147 10-line-to-4-line priority encoder
☐ 1 ADC0804 8-bit A/D converter
☐ 8 LEDs (red)
☐ 1 100-pF capacitor
☐ 8 220-Ω resistors
☐ 10 1-kΩ resistors
☐ 1 10-kΩ resistor
☐ 1 100-kΩ potentiometer (preferably a 10-turn trimmer pot)
☐ 1 SPST switch
☐ 1 package of jumper wires

BACKGROUND INFORMATION

In Experiment 20 you saw how digital-to-analog conversion takes place. In this experiment you will analyze the reverse procedure, analog-to-digital conversion. Essentially, there are three ways the latter can be achieved: with the use of *flash, counter-based,* or *successive-approximation analog-to-digital converters (ADCs).* You will examine the first and last approaches.

The flash ADC is the fastest, though potentially the most complex, ADC. Nonetheless, you will begin this experiment by building a simple but effective flash ADC using only four analog comparators. Even with just four comparators (in one IC), the basic flash A/D conversion principle is easily illustrated.

Next you will expand the flash ADC to incorporate BCD decoding. The result is a circuit that will display, in BCD format, 10 discrete voltage increments over a portion of the entire analog-input-voltage range.

Finally, you will move on to successive-approximation A/D conversion, with the use of the 20-pin ADC0804 8-bit A/D converter. You will build and then analyze operation of an ADC0804 8-bit ADC Demonstration Circuit.

PROCEDURE

Part 1: Circuit Fundamentals

Flash Analog-to-Digital Conversion

☐ 1. Flash ADCs are composed of analog comparators. Recall that such comparator outputs are either a LOW or HIGH voltage, depending on whether the voltage on one input is above or below the reference voltage on the other input. Take the comparator in Figure 21.1, for example. If the reference voltage is applied to the noninverting (+) input, and the inverting input (−) is above the 5-V reference, the comparator's output is LOW. When the inverting input drops below the 5-V reference, even by a few millivolts, the output goes HIGH.

By connecting four comparators with their inverting inputs in parallel, as shown in Figure 21.2, you can create a simple flash ADC with 4 bits of resolution. Note the five equal-value 1-kΩ resistors, R_1–R_5, connected to form a voltage divider. Since each resistor drops 1 V (remember voltage division), the noninverting inputs of comparators 1–4 are at 1, 2, 3, and 4 V, successively. As potentiometer R_6 (acting as a variable voltage divider) is varied to produce from 0 to 5 V at the input, successive comparator outputs go LOW, turning on LEDs 1–4 in turn.

For example, if the input voltage is below 1 V, all comparator outputs are HIGH and all LEDs are off. When the input voltage reaches slightly above 1 V, the output of comparator 1 goes LOW, causing LED D_1 to light. We now know that the input is at least 1 V. When the input voltage exceeds 2 V, comparator 2's output also goes LOW. LED D_2 now lights as well, indicating that at least 2 V are present at the input. With at least 3 V at the input, LED D_3 also turns on; and with at least 4 V coming in, all four LEDs are now lit.

☐   a. Referring to Figure 21.2, record missing IC pin numbers. Construct the circuit of Figure 21.2. Place a voltmeter across the input of the circuit. Apply power.

☐   b. Set R_6 for minimum input voltage. What is the voltmeter reading? ____

     What is the status of LEDs D_1–D_4? _____

☐   c. Slowly vary R_6 to increase the input voltage. As each LED, D_1–D_4, turns on in order, halt the adjustment of R_6 and record the voltage reading in

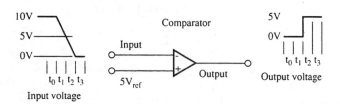

FIGURE 21.1

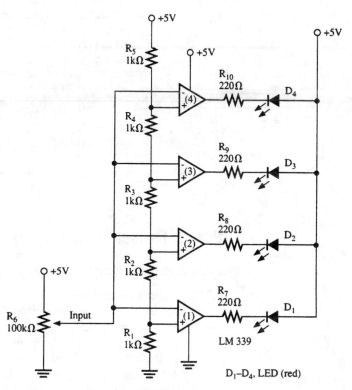

FIGURE 21.2

Table 21.1. (*Note:* Even at a low input voltage, all LEDs may glow dimly. Nevertheless, there will be a noticeable difference in LED brightness as a respective comparator output goes LOW.) Is the circuit sensitivity approximately 1 V, as expected? _____

☐ 2. Though the circuit of Figure 21.2 has a fixed resolution (1 in 4), its *sensitivity* can easily be improved. If we change R_5 to a 10-kΩ resistor, a sensitivity of 0.357 V per comparator results. (The 10-kΩ resistor drops 3.57 V. Thus only 1.43 V are left to be dropped by resistors R_1–R_4. Each resistor, R_1–R_4, in turn drops only 0.357 V.)

   ☐ a. In the circuit of Figure 21.2, replace the 1-kΩ resistor of R_5 with a 10-kΩ resistor.

   ☐ b. Set potentiometer R_6 for minimum input voltage. Then slowly adjust R_6 to increase the input voltage. As each LED, D_1–D_4, turns on in order, halt the adjustment of R_6 and record the voltage reading in Table 21.2. Is the circuit sensitivity at approximately 0.357 V, as expected? _____

TABLE 21.1

Input voltage reading	D_4	D_3	D_2	D_1
	OFF	OFF	OFF	ON
	OFF	OFF	ON	ON
	OFF	ON	ON	ON
	ON	ON	ON	ON

TABLE 21.2

Input voltage reading	D_4	D_3	D_2	D_1
	OFF	OFF	OFF	ON
	OFF	OFF	ON	ON
	OFF	ON	ON	ON
	ON	ON	ON	ON

Part 2: Further Investigation

Flash Analog-to-Digital Conversion with BCD Decoding

☐ 1. There is an obvious problem with the flash A/D converter of Figure 21.2—its outputs are not true binary or BCD. To get a BCD output, for example, you will need to decode the comparator outputs into the familiar 4-bit 0000–1001 BCD pattern. The circuit of Figure 21.3, with nine comparators and a 74147 10-line-to-4-line encoder IC (the connection diagram of which is shown in Figure A.19, Appendix A), will do the trick. As the analog voltage at the input increases, outputs from successive comparators go LOW at 0.5-V increments. The active-LOW inputs of the 74147 accept signals from the comparators and decode them for presentation in BCD form at the active-LOW outputs, where LEDs D_1–D_4 turn on to indicate the BCD number. The circuit "recognizes" input voltages in 0.5-V increments, ranging from 0.5 to 4.5 V. For example, with 2 V at the input, the outputs of comparators 1–4 are LOW, and LED D_3 is lit. If the input voltage is greater than 4.5 V, all comparator outputs are LOW, and LEDs 1 and 4 are lit.

    ☐ a. Referring to Figure 21.3, record missing IC pin numbers. Construct the circuit of Figure 21.3. Place a voltmeter at the input of the circuit. Apply power.

    ☐ b. Slowly vary potentiometer R_{11} to apply a minimum-to-maximum input voltage. Do LEDs D_1–D_4 indicate a BCD count at approximately 0.5-V

        intervals? _____ (*Note:* Even at a low input voltage, all LEDs may glow dimly. Nevertheless, there will be a noticeable difference in LED brightness as a respective comparator output goes LOW.)

Part 3: Circuit Applications

An ADC IC: ADC0804 8-Bit ADC Demonstration Circuit

☐ 1. As with DACs, ADCs come in integrated-circuit packages. Typical is the 20-pin ADC0804 successive-approximation analog-to-digital converter, the pin configuration for which is shown in Appendix A (Figure A.30). Since it is an 8-bit ADC, the voltage increment for each LSB change is found by the formula

$$V_{inc} = \frac{1}{256} \times V_{CC}$$

where V_{inc} = voltage increment

       V_{CC} = supply voltage

Thus for an 8-bit ADC,

$$V_{inc} = 0.01953 \text{ V}.$$

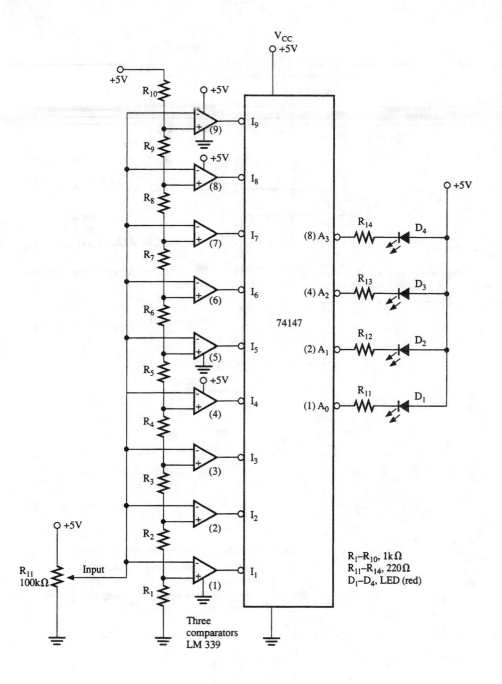

FIGURE 21.3

An ADC0804 8-bit ADC Demonstration Circuit, using the ADC0804 IC, is shown in Figure 21.4. The analog input voltage is applied to the $V_{in}(+)$ input (via potentiometer R_2). The binary outputs are taken at D_0–D_7. To test the circuit, a given analog input voltage is selected, for example, 1.567 V. The binary output is calculated by dividing 1.567 by 0.01953 (V_{inc}). The result is 80.69 in decimal, or 01010000 in binary to the nearest whole number. Switch S_1 is then opened and closed. LEDs D_5 and D_7 should be dark, the remaining LEDs lit, indicating a binary 01010000 (decimal 80). [*Note:* In this circuit a dark LED indicates a 1 (HIGH), a lit LED, a 0 (LOW).]

☐ a. Referring to Figure 21.4, record missing IC pin numbers. Construct the circuit of Figure 21.4. Place a voltmeter across the input of the circuit. Apply power.

☐ b. Using potentiometer R_2, set the input voltage at $V_{in}(+)$ to the voltage listed in step 1 of Table 21.3. Calculate and record the binary output. Record the indicated binary output appearing on LEDs D_1–D_8.

☐ c. Complete steps 2–4 as per above.

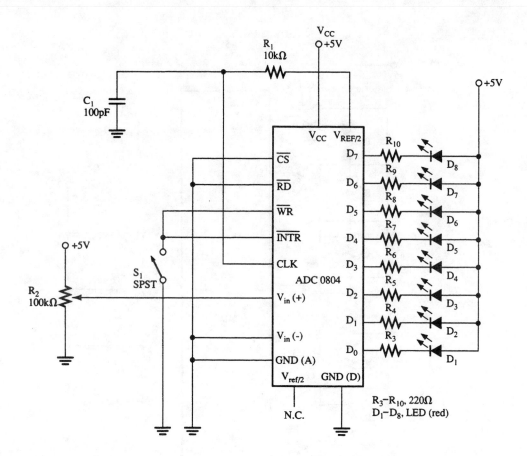

Note: A dark LED indicates a 1 (HIGH);
a lit LED, a 0 (LOW).

FIGURE 21.4

TABLE 21.3

Step	Analog voltage in	Calculated binary output	Indicated binary output
1	1.5		
2	2.0		
3	2.6		
4	3.5		
5			
6			

☐ d. In steps 5 and 6, select your own analog voltage inputs and calculate and record the binary output in each case. Record the indicated binary output appearing on LEDs D_1–D_8 as well.

SUMMARY

In this experiment you examined analog-to-digital conversion. You began by investigating a simple 4-bit flash A/D converter. Next you expanded the circuit to include BCD decoding. Finally, you built and then analyzed operation of an ADC0804 8-bit ADC Demonstration Circuit.

1. In the circuit of Figure 21.2, if resistor R_5 is 68 kΩ and resistors R_1–R_4 are 1 kΩ each, what is the reference voltage at the noninverting input of comparator 1? _____

2. In the circuit of Figure 21.2, if resistors R_1–R_5 are of equal value, and the 100-kΩ linear potentiometer R_6 is set to its midrange, which LEDs, D_1–D_4, are lit? _____

3. In the circuit of Figure 21.3, if the input voltage is 6.3 V, which LEDs, D_1–D_4, are lit? _____

4. In the circuit of Figure 21.4, if the input voltage is 3.750 V, which LEDs, D_1–D_8, are dark? _____

Writing Skills Assignment

In as few paragraphs as possible, explain how the ADC0804 8-bit ADC Demonstration Circuit works. You might first discuss what the circuit does and then how it does it.

22 RAM (Random-Access Memory)

OBJECTIVES

After completing this experiment, you will be able to

- Write data into and read data from a RAM chip
- Verify operation of a TTL 7489 16-×-4-bit RAM
- Build and analyze operation of a Programmable Light Sequencer Circuit using the 7489 RAM IC

REFERENCE READING

Review Ronald Reis, *Digital Electronics Through Project Analysis,* Chapter 13, Section 13.2.

**EQUIPMENT &
MATERIALS NEEDED**

Equipment

☐ 1   5-V power supply
☐ 1   555 clock generator
☐ 1   solderless circuit board

Materials

☐   1   7489 64-bit (16 × 4) RAM
☐   1   7493 binary counter
☐   1   74154 4-line-to-16-line decoder
☐ 20   LEDs (red)
☐   6   220-Ω resistors
☐   1   330-Ω resistor
☐   5   1-kΩ resistors
☐   2   N.O. push-button switches
☐   1   SPST switch
☐   1   8-position DIP switch (or equivalent)
☐   1   package of jumper wires

Semiconductor memories are integrated circuits that store 0s and 1s in the form of LOW and HIGH voltages. There are basically two types: RAM (random-access memory), also known as read/write memory, and ROM (read-only memory). The former can be further divided into static and dynamic RAM. Static RAM is the subject of this experiment.

A typical TTL RAM chip of the type you will be using in this experiment is shown in Figure 22.1. Note that it contains the following:

1. A pin for V_{CC} and a pin for ground.
2. Four address pins; thus 16 slots can be addressed.
3. Four data-input pins; thus the chip will input 1 nibble (4 bits) at a time.
4. Four data-output pins; thus the chip will read out 1 nibble at a time.
5. A read/write ($R\overline{W}$) pin. When this pin is HIGH, data is read out. When the pin is LOW, data is written in.
6. A chip enable (or select) pin. When this pin is HIGH, the data outputs go to an open-circuit (high-Z) condition. When the pin is LOW, the chip is enabled.

In this experiment you will work with the TTL 7489 64-bit (16 × 4) RAM, the block diagram and memory matrix chart for which are shown in Figure 22.2. The diagram is similar to that of the typical TTL RAM chip of Figure 22.1, except in one important respect. Observe that the 7489 IC has bubbles on the data-out lines. Thus the data out will be the *complement* of the data put in. If, for instance, you place a 0101 nibble at memory location 0000, you will read out its complement from that location, 1010. (See the memory matrix chart in Figure 22.2.)

In this experiment you will first test the 7489 RAM to see how it performs a write and a read operation. Next you will check all 16 address locations by storing a nibble in each location, then retrieving its complement. Finally, you will build and then analyze operation of a Programmable Light Sequencer Circuit that uses the 7489 RAM chip.

FIGURE 22.1

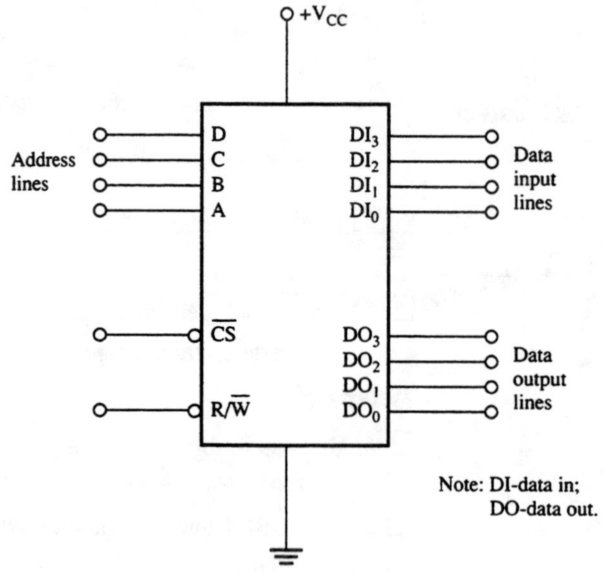

Note: DI-data in;
DO-data out.

Part 1: Circuit Fundamentals

Writing and Reading Data into RAM

☐ 1. To confirm 7489 RAM operation, you can build the RAM test circuit of Figure 22.3. Note first that the chip select ($\overline{CS}$) pin is grounded. Thus the IC is enabled. Second, all address pins, A_0–A_3, are also grounded. As a result, only address 0000 is selected. Third, data-input lines, D_1–D_4, can be individually brought LOW or HIGH. Fourth, the $\overline{WE}$ pin is normally held HIGH through a 1-kΩ resistor. Thus the chip is normally in the read mode. When switch S_5 is pressed, however, data is written into address 0000. Finally, data-output lines, Q_1–Q_4, are

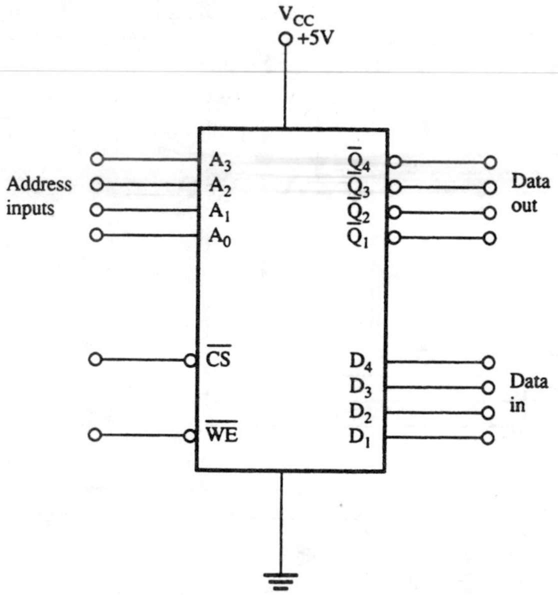

	A_3	A_2	A_1	A_0	D_4	D_3	D_2	D_1	$\overline{Q_4}$	$\overline{Q_3}$	$\overline{Q_2}$	$\overline{Q_1}$
Example	0	0	0	0	0	1	0	1	1	0	1	0
	0	0	0	1								
	0	0	1	0								
	0	0	1	1								
	0	1	0	0								
	0	1	0	1								
	0	1	1	0								
	0	1	1	1								
	1	0	0	0								
	1	0	0	1								
	1	0	1	0								
	1	0	1	1								
	1	1	0	0								
	1	1	0	1								
	1	1	1	0								
	1	1	1	1								

FIGURE 22.2

connected to the cathodes of individual LEDs. Hence, when a data-output line goes LOW, an LED lights. When it goes HIGH, its respective LED remains off.

☐ a. Referring to Figure 22.3, record missing IC pin numbers. Construct the circuit of Figure 22.3. Close switches S_1–S_4. Apply power.

☐ b. Place the binary number 0101 on the data-in lines by opening and closing appropriate switches S_1–S_4.

☐ c. Press and release switch S_5.

☐ d. What is the status of LEDs D_1–D_4? ________________ Are the LEDs indicating the complement of the nibble stored in address location 0000? (Remember, an on LED indicates a LOW, an off LED, a HIGH.)

____

☐ e. Remove, then reapply, power. What is the status of LEDs D_1–D_4? Why is this so? (Remember, this is a volatile memory.) ______________

____

____

____

Part 2: Further Investigation

Checking Out the 7489 RAM IC

☐ 1. You now need to check out all memory locations in the 7489 IC.

☐ a. Expand the circuit of Figure 22.3 as shown in Figure 22.4. (Note that the address lines are no longer grounded, but can now be brought LOW or HIGH through DIP switches S_5–S_8.) Apply power to the circuit.

☐ b. Referring to Table 22.1, place the indicated data, D_1–D_4, in address locations 0000–1111 (steps 1–16). The procedure for each step is to (1) select the appropriate address with switches S_1–S_4; (2) select the appropriate input data with switches S_5–S_8; (3) press S_5 to write the data into memory; (4) release S_5 to read the data out from memory; and (5) record the output data in the space provided.

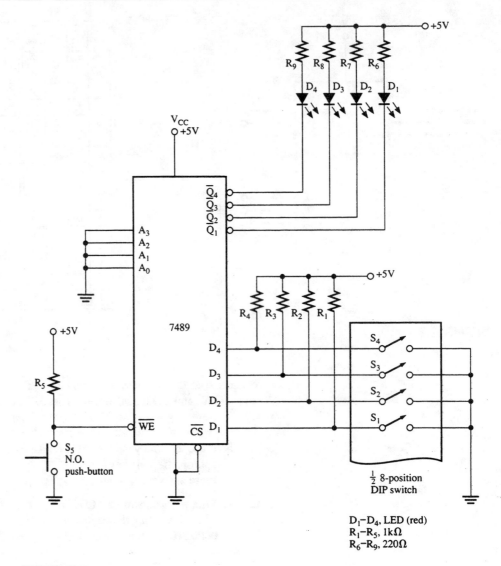

FIGURE 22.3

D₁–D₄, LED (red)
R₁–R₅, 1kΩ
R₆–R₉, 220Ω

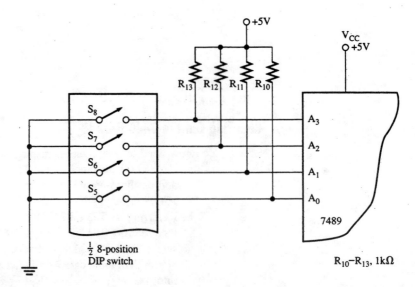

R₁₀–R₁₃, 1kΩ

FIGURE 22.4

Step	A_3	A_2	A_1	A_0	D_4	D_3	D_2	D_1	$\bar{Q}_4$	$\bar{Q}_3$	$\bar{Q}_2$	$\bar{Q}_1$
1	0	0	0	0	1	1	1	1				
2	0	0	0	1	1	1	1	0				
3	0	0	1	0	1	1	0	0				
4	0	0	1	1	1	0	0	0				
5	0	1	0	0	0	0	0	0				
6	0	1	0	1	0	1	1	0				
7	0	1	1	0	0	1	1	0				
8	0	1	1	1	1	0	0	1				
9	1	0	0	0	1	1	1	0				
10	1	0	0	1	1	1	0	1				
11	1	0	1	0	1	0	1	0				
12	1	0	1	1	0	1	0	1				
13	1	1	0	0	1	1	1	0				
14	1	1	0	1	0	1	1	0				
15	1	1	1	0	1	1	1	1				
16	1	1	1	1	0	0	0	1				

TABLE 22.1

☐ c. When all memory slots have been filled with the data indicated, select any address at random and read the data from memory. Is the output data the complement of the input data stored? _____

☐ d. Repeat the above procedure a few times. Summarize your conclusions with regard to Part 2:

_____

_____

_____

Part 3: Circuit Applications

The 7489 at Work: Programmable Light Sequencer Circuit

☐ 1. Putting the 7489 to work is easy—dozens of circuit applications exist, one of which is shown in Figure 22.5. Consisting of a clock generator, a 7493 binary counter, a 7489 RAM, and a 74154 4-line-to-16-line decoder, this Programmable Light Sequencer Circuit will turn on a series of 16 LEDs in any *programmed* sequence. Though the LEDs, D_0–D_{15}, are hardwired into the circuit, the sequence in which they are turned on, one at a time, is determined by the user and can be changed simply by reprogramming the memory chip—no rewiring is necessary.

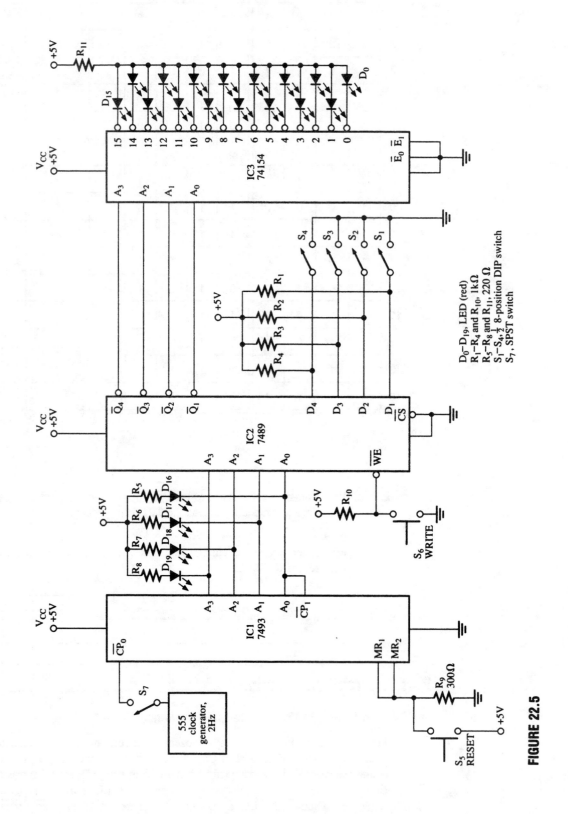

FIGURE 22.5

Operation of the circuit is straightforward. Clock pulses, via switch S_7, are sent to IC1, which acts as an address selector for IC2, the 7489 RAM. LEDs D_{16}–D_{19} give a binary indication of the address selected—with an off LED indicating a HIGH and an on LED indicating a LOW. DIP switches S_1–S_4 are used to place a nibble on the data-in lines, D_1–D_4. Switch S_5 is the reset for IC1; switch S_6 is the write enable for IC2. IC3, a 74154 4-line-to-16 line decoder, receives the complement of a nibble stored in the 7489 RAM, decodes it, and brings a respective output line LOW, turning on the given LED. The LED will remain on until a new clock pulse is generated, causing the next address of IC2 to be selected. If that address contains a different nibble, a new LED (D_0–D_{15}) will turn on.

To more fully understand circuit operation, let's go through a couple of programming sequences. Referring to Example 1 in Table 22.2, suppose you want LED D_7 to come on with the first clock pulse when the circuit is running. That means the first of 16 memory slots of IC1, at address 0000, must contain the 4-bit nibble that when presented to IC3 will be decoded to turn on LED D_7. In other words, the first memory slot of the 7489 RAM must contain a 1000. Why a 1000 and not a 0111? Remember, the 7489 RAM reads out the complement of the data stored. If you store a 1000, a 0111 (7) will be read out to the 74154 IC

	Address (7489)				LED selected (74154)	Binary address of LED (true binary number) (74154)				Complement of true binary number (7489)			
	A_3	A_2	A_1	A_0	D_x	D	C	B	A	(S4) D_4	(S3) D_3	(S2) D_2	(S1) D_1
Example 1	0	0	0	0	07	0	1	1	1	1	0	0	0
Example 2	0	0	0	1	011	1	0	1	1	0	1	0	0

TABLE 22.2

when address 0000 on the 7489 IC is selected. IC3 then decodes the 0111 nibble to turn on LED D_7.

Suppose, with the second clock pulse, you wish to turn on LED D_{11}. That means address 0001 of IC2 will contain the nibble that when read out and sent to IC3 will be decoded to turn on LED D_{11}. The stored nibble is 0100; the nibble read out is 1011, or a decimal 11.

The actual programming procedure is as follows:

- Close S_7, allowing clock pulses to arrive at the clock input ($\overline{CP_0}$) of IC1. Wait for address LEDs D_{16}–D_{19} to indicate a 0000. (Remember, an off LED indicates a HIGH, an on LED, a LOW. Thus, for address 0000, all four LEDs will be on.) When address 0000 is indicated, quickly open S_7, preventing further pulses from arriving at IC1.
- Determine the LED you wish to turn on.
- Determine the true binary number that will select the LED you wish turned on.
- Enter the complement of the true binary number using DIP switches S_1–S_4.
- Press and release S_6, causing the number to be written into the first memory location. When S_6 is released, the complement of the binary number stored (the true binary number) is read out and the chosen LED will light.
- To proceed to the next address location, 0001, again close S_7 and wait for a new pulse to arrive on IC1. When the address LEDs indicate 0001 (on, on, on, off, from left to right), open S_7. Now repeat the sequence, determining the true binary number of the LED you wish to turn on, entering its complement.

☐ a. Referring to Figure 22.5, record missing IC pin numbers. Construct the circuit of Figure 22.5. Close switches S_1–S_4 and open switch S_7. Apply power.

☐ b. Following the steps outlined above, enter the data necessary to turn on LEDs D_7 and D_{11} in sequence as indicated in Table 22.2 (Examples 1 and 2). When you have finished, allow the circuit to run through all memory locations a few times. Observe if LEDs D_7 and D_{11} come on when addresses 0000 and 0001, respectively, are selected. (*Note:* Other LEDs may come on as other addresses are selected, since "garbage"—random nibbles—exists in these unprogrammed locations.)

☐ c. Now proceed to program the circuit to turn on some or all of the remaining LEDs in any sequence desired. If you want a given LED to remain on for two (or more) pulses, just program the same nibble in successive address locations. Before actually programming, however, complete Table 22.2 with the appropriate data.

☐ d. Summarize your conclusions with regard to Part 3:

_____

_____

_____

SUMMARY

In this experiment you explored operation of the TTL 7489 64-bit (16 × 4) RAM chip. You began by verifying its read/write capability. Next you validated each memory location. Finally, you built and then analyzed operation of a Programmable Light Sequencer Circuit that uses the 7489 IC.

REVIEW QUESTIONS

1. The 7489 RAM has _____ memory slots, each holding one _____ (4 bits).

2. Explain the function of the $\overline{WE}$ pin on the 7489 RAM chip. _____

_____

_____

3. Referring to Figure 22.4, if a 0110 nibble is placed in address location 1111, which

   LEDs will light when the data is read out? _____
4. Outline the steps necessary to program a 7489 IC.

   _____

   _____

   _____

   _____

   _____

   _____

Writing Skills Assignment

In as few paragraphs as possible, explain how the Programmable Light Sequencer Circuit works. You might first discuss what the circuit does and then discuss how it does it.

A IC Pin Diagrams

Below are the IC pin diagrams for the integrated circuits used in this laboratory manual. You will notice throughout that there are a number of ways to designate the same pin characteristics. For example, a clock input may be seen as CLK, CP_0, or just C. The enable function might be shown with an E or a G. And binary numbers (addresses, data in, and data out) are indicated in numerous ways, some of which are listed below:

8	4	2	1
D	C	B	A
S_3	S_2	S_1	S_0
S_4	S_3	S_2	S_1
D_3	D_2	D_1	D_0
D_4	D_3	D_2	D_1
A_3	A_2	A_1	A_0
A_4	A_3	A_2	A_1
Q_3	Q_2	Q_1	Q_0
Q_4	Q_3	Q_2	Q_1

The designation used depends on the IC manufacturer; each company has its own preference. As an electronics technician, you will want to become familiar with each method.

FIGURE A.1

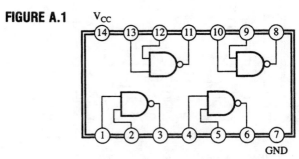

7400 quad two-input NAND gate

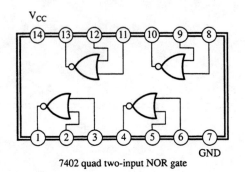

7402 quad two-input NOR gate

FIGURE A.2

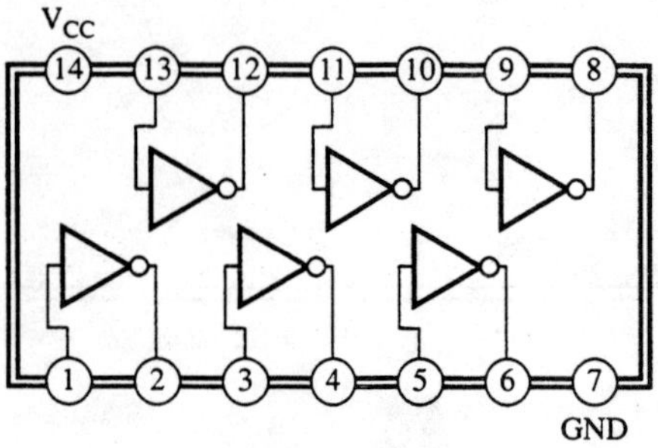

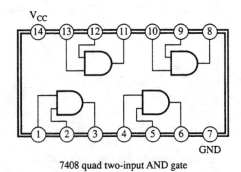

7404 hex inverter

FIGURE A.3

7408 quad two-input AND gate

FIGURE A.4

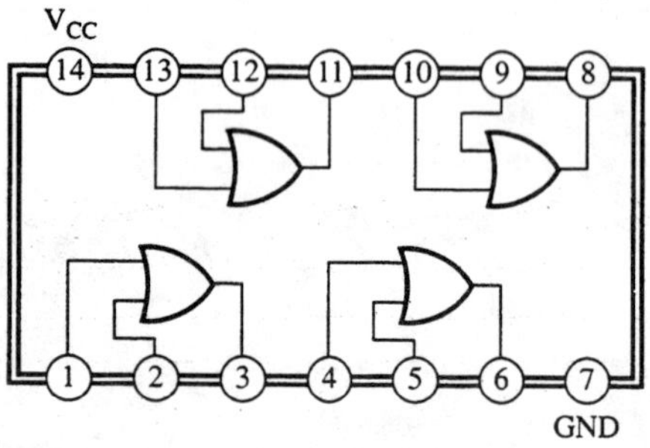

7432 quad two-input OR gate

FIGURE A.5

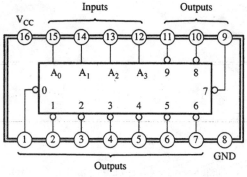

7442 BCD-to-decimal decoder

FIGURE A.6

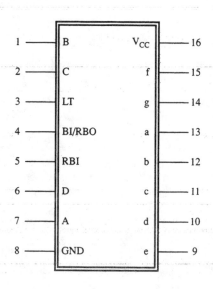

7447 BCD-to-seven-segment decoder/driver

FIGURE A.7

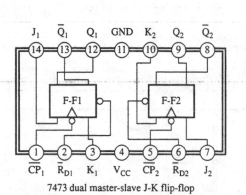

7473 dual master-slave J-K flip-flop

FIGURE A.8

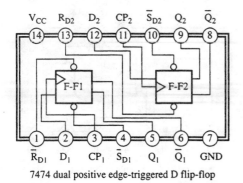

V_{CC} R_{D2} D_2 CP_2 $\overline{S}_{D2}$ Q_2 $\overline{Q}_2$

$\overline{R}_{D1}$ D_1 CP_1 $\overline{S}_{D1}$ Q_1 $\overline{Q}_1$ GND

7474 dual positive edge-triggered D flip-flop

FIGURE A.9

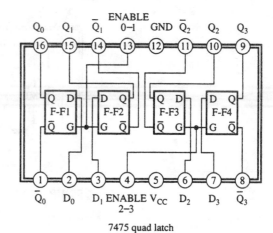

Q_0 Q_1 $\overline{Q}_1$ ENABLE 0–1 GND $\overline{Q}_2$ Q_2 Q_3

$\overline{Q}_0$ D_0 D_1 ENABLE V_{CC} D_2 D_3 $\overline{Q}_3$
2–3

7475 quad latch

FIGURE A.10

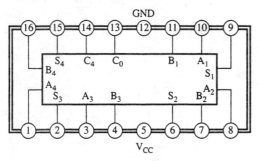

GND

S_4 C_4 C_0 B_1 A_1
B_4 S_1
A_4 A_2
S_3 A_3 B_3 S_2 B_2

V_{CC}

7483 4-bit full adder

FIGURE A.11

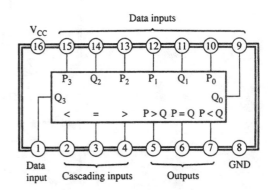

Data inputs

V_{CC}

P_3 Q_2 P_2 P_1 Q_1 P_0
Q_3 Q_0
< = > P > Q P = Q P < Q

Data input   Cascading inputs   Outputs   GND

7485 4-bit magnitude comparator

FIGURE A.12

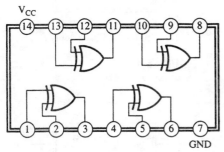

V_{CC}

GND

7486 quad two-input exclusive-OR gate

FIGURE A.13

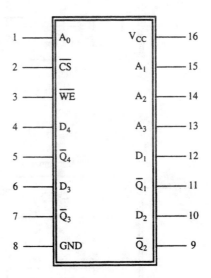

1	A_0	V_{CC}	16
2	$\overline{CS}$	A_1	15
3	$\overline{WE}$	A_2	14
4	D_4	A_3	13
5	$\overline{Q}_4$	D_1	12
6	D_3	$\overline{Q}_1$	11
7	$\overline{Q}_3$	D_2	10
8	GND	$\overline{Q}_2$	9

7489 64-bit (16 × 4) RAM

FIGURE A.14

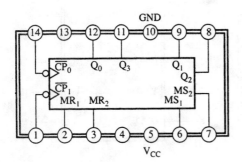

GND

$\overline{CP}_0$ Q_0 Q_3 Q_1
Q_2
$\overline{CP}_1$ MS_2
MR_1 MR_2 MS_1

V_{CC}

7490 decade counter

FIGURE A.15

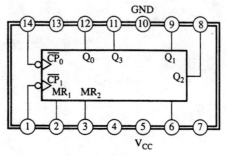

7493 binary counter

FIGURE A.16

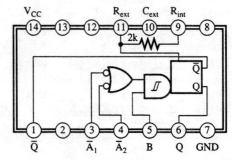

74121 monostable multivibrator

FIGURE A.17

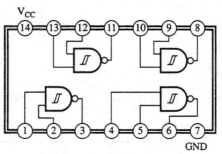

74LS132 quad two-input NAND Schmitt trigger

FIGURE A.18

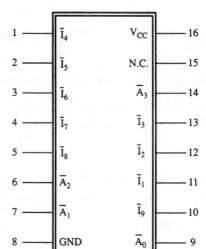

74147 10-line-to-4-line priority encoder

FIGURE A.19

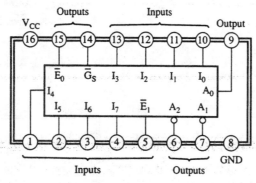

74148 priority encoder

FIGURE A.20

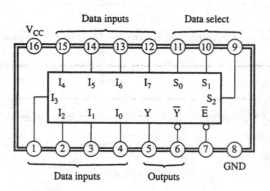

74151 data selector multiplexer

FIGURE A.21

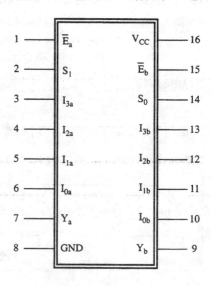

74153 dual 4-line-to-1-line multiplexer

FIGURE A.22

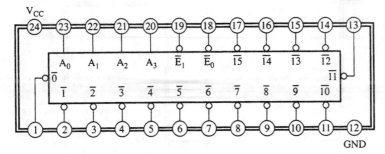

74154 4-line-to-16-line decoder/demultiplexer

FIGURE A.23

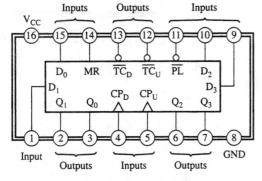

74193 synchronous up-down 4-bit binary counter

FIGURE A.25

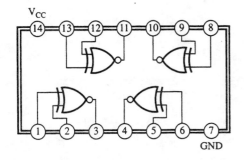

74LS266 quad two-input exclusive-NOR gate

FIGURE A.27

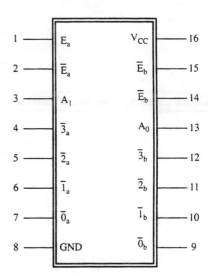

74155 2-line-to-4-line demultiplexer

FIGURE A.24

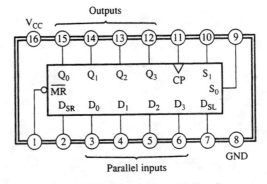

74194 4-bit bidirectional universal shift register

FIGURE A.26

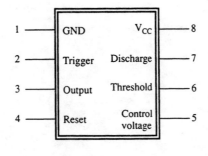

555 timer

FIGURE A.28

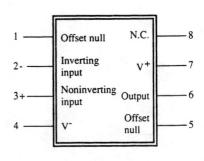

741 operational amplifier

FIGURE A.29

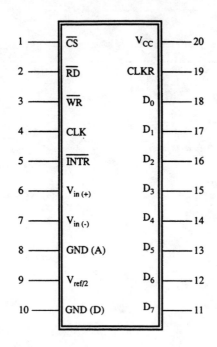

ADC0804 8-bit A/D converter

FIGURE A.30

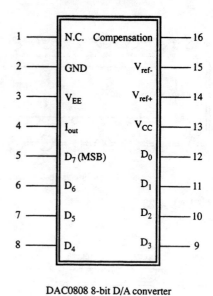

DAC0808 8-bit D/A converter

FIGURE A.31

B Data Manual Sources

Following are the names, addresses, and telephone numbers of 15 companies that supply data manuals for their semiconductor products, often free for the asking:

Cypress Semiconductor
23586 Calabasas Road, Suite 201
Calabasas, California 91302
(818) 884-7800

Dallas Semiconductor
4350 Beltwood Parkway South
Dallas, Texas 75244-3219
(214) 450-0400

Harris Semiconductor Literature Dept.
P.O. Box 883, MS 53-035
Melbourne, Florida 32901
(407) 724-7400

Hewlett-Packard
640 Page Mill Road
Palo Alto, California 94304
(415) 857-1501

Intel Corporation
3065 Bowers Avenue
Santa Clara, California 95051
(800) 548-4725

Intersil, Inc.
10600 Ridge View Court
Cupertino, California 95014
(408) 996-5000

Motorola Literature Distribution
P.O. Box 20912
Phoenix, Arizona 85036
(602) 344-7100

National Semiconductor Corporation
2900 Semiconductor Drive
Santa Clara, California 95051
(408) 737-5000

NEC Electronics, Inc.
401 Ellis Street
P.O. Box 7241
Mountain View, California 94039
(415) 960-6000

RCA Solid State Division
Box 3200
Somerville, New Jersey 08876
(609) 338-5042

Saratoga Semiconductor
686 West Maude Avenue
Sunnyvale, California 94086
(408) 522-7500

Signetics Company
811 East Arques Avenue
P.O. Box 3409
Sunnyvale, California 94088-3409
(408) 991-2000

Texas Instruments
P.O. Box 2909
Austin, Texas 78769
(512) 680-5082

Waferscale Integration, Inc.
47280 Kato Road
Fremont, California 94538
(415) 656-5400

Xicor, Inc.
851 Buckeye Court
Milpitas, California 95035
(408) 432-8888

C Troubleshooting: A 10-Point Checklist

If a circuit fails to operate or operates improperly, troubleshooting is required. While complex circuits often demand specific troubleshooting procedures and techniques, the 10-point checklist below should enable you to locate the source of most problems in the circuits created in this manual. Look the list over carefully now and refer to it again if troubleshooting should be necessary.

1. Are all wiring connections correct and complete? In other words, do all wires go where they are supposed to and do they make a complete connection? (That is, be sure that none are broken or open.)
2. Are all IC pins fully inserted into the solderless circuit board? It is very easy for a pin to get tucked under the IC body when the chip is being installed. If you look at it from the side, the pin seems to be inserted correctly. In reality, it is not entering a hole and is thus not making contact with the circuit.
3. Are all ICs receiving power? Every digital IC has at least one positive and one ground pin. Make sure these pins are receiving the correct voltages.
4. Are V_{CC} and V_{DD} within range? For TTL, V_{CC} must not exceed 5 V. For CMOS, V_{DD} must not exceed 15 V.
5. Do all inputs go somewhere? Check that this is true for both used and unused gates.
6. Are all polarized components (ICs, diodes, transistors, etc.) inserted in the correct direction?
7. In a TTL circuit, are 0.1-μF decoupling capacitors placed close to and across the supply pins of every fourth chip? Such decoupling capacitors will help to stabilize an otherwise erratic circuit.
8. Are the ICs running hot? Give them the "touch test" by placing your finger on the IC case. If you can keep your finger in place, the chip is probably OK. If you cannot, the IC, especially if it's a CMOS chip, is probably burned out. It will have to be replaced.
9. Are all inputs and outputs of all ICs properly interfaced?
10. Are all soldering connections good? They should be shiny rather than dull gray.

D Needed Equipment You Can Build: A 5- and 9-V Power Supply and a 555 Clock Generator

Below you will find the schematics for two pieces of test equipment that you can easily and quickly build. The 5- and 9-V Power Supply (Figure D.1) is fully regulated and ideal for digital circuit experimentation. The components are readily available and can be purchased for under $10 at most electronics stores.

The 555 Clock Generator is also easy and inexpensive to build (Figure D.2). The output frequencies indicated in the accompanying chart are approximate, but close enough to those called for in the experiments.

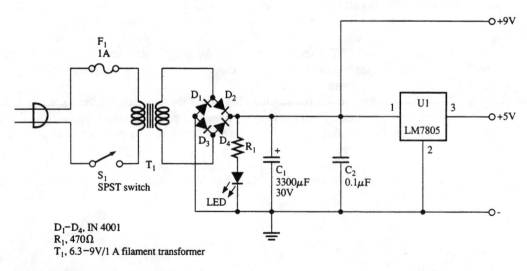

D_1–D_4, IN 4001
R_1, 470 Ω
T_1, 6.3–9V/1 A filament transformer

FIGURE D.1 The 5- and 9-V Power Supply

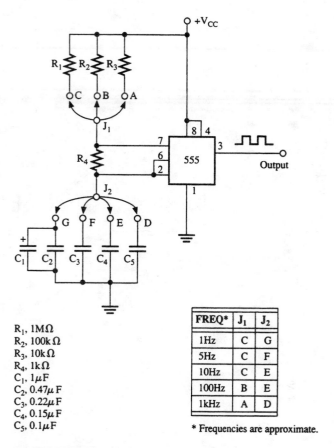

R₁, 1MΩ
R₂, 100kΩ
R₃, 10kΩ
R₄, 1kΩ
C₁, 1μF
C₂, 0.47μF
C₃, 0.22μF
C₄, 0.15μF
C₅, 0.1μF

FREQ*	J_1	J_2
1Hz	C	G
5Hz	C	F
10Hz	C	E
100Hz	B	E
1kHz	A	D

* Frequencies are approximate.

FIGURE D.2   The 555 Clock Generator